Impacts of Industrial Robotics

The Economics of Technological Change
Edwin Mansfield, *General Editor*

Impacts of Industrial Robotics

Potential Effects on Labor and Costs
within the Metalworking Industries

Steven M. Miller

THE UNIVERSITY OF WISCONSIN PRESS

The University of Wisconsin Press
114 North Murray Street
Madison, Wisconsin 53715

5   4   3   2   1

Printed in the United States of America

Library of Congress Cataloging-in-Publication Data
Miller, Steven M.
    Impacts of industrial robotics.
    (The Economics of technological change)
    Bibliography: pp. 243–249.
    Includes index.
    1. Metal-work—Automation—Economic aspects.
2. Metal-workers—Effect of technological innovations on.
3. Robots, Industrial. I. Title. II. Series.
HD9506.A2M527   1988   338.4'5671   87-40151
ISBN 0-299-10500-8
ISBN 0-299-10504-0 (pbk.)

To my parents, Phyllis and David Miller,
who have always been supportive and encouraging

Contents

Figures

Tables

Foreword

According to leading expert opinion, industrial robots are among the most important innovations of recent decades. Yet there is very considerable uncertainty (and disagreement) as to the rate at which the use of robots will spread and what their effects on employment, costs, and market structure will be. As stressed in 1983 by Congress's Office of Technology Assessment: "A central question for an analysis of the social and economic impacts of programmable automation is whether programmable automation is likely to spread especially rapidly among firms and industries, and why."

In this timely and interesting book, Steven Miller analyzes the job displacement effects of industrial robots, as well as their potential impacts on production costs. Combining engineering and economic data, he concludes that robot use will continue to be concentrated within large firms for the next several years, displacement impacts will be concentrated within particular geographic locations and industries, and that very few workers will actually lose their employment as a direct result of robot use. He also analyzes the effects of flexible manufacturing systems on production costs.

This is the fourth volume in the University of Wisconsin Press Series on the Economics of Technological Change. Morris Teubal's *Innovation Performance, Learning, and Government Policy: Selected Essays* dealt with factors influencing success and failure in innovation and with government technology policy. Stanislaw Gomulka's *Growth, Innovation and Reform in Eastern Europe* was concerned with fundamental questions concerning the effects of economic systems on the rate of innovation. Franco Malerba's *The Semiconductor Business: The Economics of Rapid Growth and Decline* analyzed the decline of the European semiconductor industry, relative to that in the United States and Japan. All of these books have attempted to understand why some organizations are more successful than others in promoting and accepting new technologies, a question that will continue to be a central concern of future books in this series.

Director, Center for   EDWIN MANSFIELD
Economics and Technology
University of Pennsylvania

Preface

In 1978, my first year on the Carnegie-Mellon campus, there was essentially no research activity in the area of discrete-product manufacturing. I was not even aware of the absence of such activity, since I had no previous exposure to the topic and certainly no plans for studying it. During the next several years, though, interest in industrial automation and manufacturing operations suddenly swelled on campus as well as across the country. Once the university's Robotics Institute was formed in 1980, interest in manufacturing seemed to explode on campus. Through the Robotics Institute, substantial resources were made available for computer scientists and engineers to work together in marrying advanced computer technology with methods of industrial production. The goal was to create more flexible and more intelligent systems of automation and control. Resources were also provided to faculty interested in technology assessment, sociology, and management in order to define and study existing and emerging impacts of the use of robotics. Industrial sponsors of the Institute, most notably Westinghouse, provided opportunities for researchers to learn about and work on real-world problems in factories. Also, the school's administration propagated the attitude that this type of work was important. As a result of these resources, opportunities, and incentives, many faculty and graduate students—most of whom, like myself, had no previous interest or background in manufacturing—found themselves working intensively on problems related to the development and use of new technology for factory automation.

I was in the doctoral program in engineering and public policy, a department in the engineering college that addresses policy-related research problems for which technical issues cannot be treated as a "black box." I was working with Professor Robert U. Ayres, whose main interest at that time was the assessment of the environmental and economic impact of changes in energy technologies. When we first met, he, too, had no particular interest in discrete-product manufacturing. But in 1979, Professor Ayres and I met Paul Wright, a newly arrived professor in the Mechanical Engineering Department. Professor Wright studied cutting tool materials used in metalworking operations and, as a result, had a general knowledge of discrete-product manufacturing. He showed us publications that discussed both the need for flexibly automated machining systems for batch production of metal parts and the attempts to build such systems. For both Professor Ayres and myself, this was our first introduction to manufacturing technology within the metalworking industries. We frequently discussed with Professor Wright the issue of how the value of increased flexibility in discrete-product manufacturing could be defined and measured. Professor Ayres was interested in analyzing the economic impact of the emerging types of factory automation, while Professor Wright thought such discussions might provide some insight into the type of system he should actually try to develop in his laboratory.

In the spring of 1980, Professor Raj Reddy, the director of the recently formed Robotics Institute, commissioned Professor Ayres to do a technology assessment on the potential applications and implications of robotics. Then, in 1981, Professor Ayres and I played the lead role in organizing and directing a project course with nearly 30 students which produced a report entitled "The Impact of Robots on the Workforce and Workplace." Material from the technology assessment and the student project eventually evolved into the book *Robotics: Applications and Social Implications* (1983) by Professor Ayres and myself. Professor Ayres and I believed that there would be two major economic benefits of increased robot use in manufacturing. The most obvious benefit would be a reduction in direct labor costs as a result of production worker displacement from current jobs. The less obvious, but more important, benefit would be improvement in the utilization of capital equipment, especially in those industries which produced discrete products in batches. Thinking about a potential reduction in labor costs clearly suggested the need for the analysis of another type of impact of robot use, the potential effect on job displacement. This led us to look in detail at the occupations that robots are well suited to perform and the number of people performing these occupations throughout industry. The 1983 book summarized our first-round analysis of the potential effect of robot use on job displacement and cost reduction.

These were the two impacts I chose to focus on in more depth for my doctoral dissertation, "Potential Impacts of Robotics on Manufacturing Costs within the Metalworking Industries," completed in mid-1983. One concern I encountered when I began the dissertation was that the well-publicized Ayres and Miller estimate of the number of workers that could *potentially* be displaced by robot use was nearly 10 times higher than other estimates of displacement based on projections of the number of robots that would actually be sold. My own analysis of the displacement issue began by questioning if there were reason to doubt our estimated theoretical potential for substituting robots for human workers. It continued by developing a plausible explanation for the large difference between the Ayres and Miller estimate of the potential opportunity for using robots compared with the number of robots forecasted to be used in the near future. A related issue was further exploration of whether or not a public policy problem would result if our estimated potential displacement, which we viewed as an upper bound, were to be realized. In a highly simplified way, my analysis also examined whether or not the level of displacement given by our upper-bound estimates could likely be offset by attrition or output increases.

Another concern encountered was that in our earlier work (Ayres and Miller, 1983) we lacked a way of estimating the magnitude of cost savings that could be realized by narrowing the efficiency gap between batch and mass production. Ideally, we wanted to know the distribution of value added by custom, batch, and mass modes of production across all of the industries in the metalworking sector. Then we wanted to estimate the unit cost differential between batch production and mass production. The magnitude of this differential would highlight the penalty costs incurred in producing customized products with the current generation of technology. This cost differential, taken in combination with the distribution of value added by batch size, would emphasize the value of improving the efficiency of batch-production operations. Finally, we wanted to estimate the amount by which cost per unit in custom and batch production could be reduced if robotics, in conjunction with other computer-aided manufacturing technologies, could bring low- and medium-volume production closer in efficiency to mass production. We felt that if we could do this, we could tangibly demonstrate the potential economic benefit realizable by accelerating the use of robotics and other flexible forms of automation.

My own analysis of the potential for cost reduction focused on constructing a measure that could be used to make comparisons of unit cost *across* industries in the metalworking sector. The measure is the ratio of value added to the pounds of material processed by each industry. This ratio, interpreted as the production cost per unit of material processed, is used as a surrogate measure of cost per unit of output produced. The idea of using this type of measure as a means of comparing cost across industries had been suggested by Ayres in our previous work. Given a means of comparing cost across industries, along with other data that allowed me to estimate the potential for increasing capacity utilization and throughput in low- and medium-volume production, I was able to generate empirically based estimates of the three things we had set out to show earlier: the distribution of value added by custom, batch, and mass modes of production; the unit cost differential between batch and mass production; and the reduction in unit cost realizable by bringing batch production closer in efficiency to mass production.

This book is a substantially revised version of my 1983 doctoral dissertation on the impact of robotics. The topics analyzed in the dissertation, which are mentioned above, are the same ones presented here. However, the current presentation has been reorganized, shortened in some sections, lengthened in others, and revised and updated with recent findings that have appeared in the literature between 1983 and 1986. In reworking the manuscript, I have benefited a great deal from the input of a number of people. The comments of an anonymous referee, working through the University of Wisconsin Press, were especially helpful. Dr. Timothy Hunt, of the W. E. Upjohn Institute for Employment Research in Kalamazoo, Michigan, provided feedback on the discussion of job displacement. Over the years, despite differences in our perspectives, Dr. Hunt has been willing to debate issues with me, which has been most helpful and greatly appreciated. Also, Professor Faye Duchin of New York University provided useful comments on how to improve parts of the analysis in the original dissertation manuscript. Professor Alan Porter of the Georgia Institute of Technology and Peter Heytler of the Industrial Development Institute of the University of Michigan also contributed by taking time to check parts of the manuscript.

There are people who contributed to the revised work by supplying important data or information. Most notable is a market analyst from a major robot supplier who supplied actual data on price quotations for robot applications. Ron Potter of Advanced Manufacturing Systems in Norcrosse, Georgia, provided background on the total cost of installing robot systems. Also, Tony Urrico, a graduate student in the Department of Engineering and Public Policy, contributed by updating and revising a number of calculations.

Professor Edwin Mansfield, editor of this series on technological change, deserves special thanks for encouraging me to turn the dissertation into a book. The people at the University of Wisconsin Press deserve thanks for their neverending patience, their encouragement, and their editorial assistance. The Graduate School of Industrial Administration at Carnegie-Mellon provided general support during the period of reworking the manuscript. GSIA also helped me to establish working relationships with several manufacturing companies during the past several years. While this more recent work does not show up directly in the manuscript, it has given me a more realistic sense of the impact of technological change on the factory, and this has helped me keep things in perspective.

The key people who helped me complete the doctoral dissertation upon which this book is based deserve special acknowledgement. For the five years that we worked together, Professor Robert Ayres provided guidance on how to work in the largely unexplored area that lies between emerging technological capabilities and impending societal implications. He provided examples of how creativity and imagination can help establish a starting point for exploring new problems. Dr. Larry Westphal, formerly of the Economics of Industry Division of the World Bank, Washington, D.C., provided useful criticism and helpful suggestions. Financial support during the 1980–83 period was provided by the Department of Engineering and Public Policy and by the Westinghouse Electric Corporation through its grant to the Social Impacts of Information and Robotics Technology Program at Carnegie-Mellon.

And finally, my most important acknowledgement is to my wife, Dr. Patricia Meyer. She has helped in every phase of this project, from the proofreading of my dissertation several years ago, to the revision, editing, and production of the completed book.

Pittsburgh, Pennsylvania   S. M.

Impacts of Industrial Robotics

1

Motivation

This book addresses two issues concerning the impact of robotics on manufacturing labor and costs. The first issue is the extent to which manufacturing workers will be displaced from their current jobs. The second issue is the extent to which manufacturing costs can be reduced by increasing capital utilization and throughput, as well as by reducing direct labor costs. In the analysis of job displacement, the technological perspective is narrowly confined to the impact of using robotic manipulators that are "retrofitted" into existing production facilities. In the analysis of the potential for cost reduction, the technological perspective is broader and considers the impact of integrating robots with other types of computer-aided manufacturing technology into flexible manufacturing systems.

The material is generally directed at people interested in studying the social and economic impacts of technological change. More specifically, it is directed at those concerned with the impact of new production technologies in manufacturing industries. The book will be especially relevant to people interested in a selected group of manufacturing industries referred to as either the metalworking sector, the engineering sector, or the durable goods sector. Specifically, this group includes the industries in the following major groups identified in the Standard Industrial Classification (SIC) code: Fabricated Metal Products (SIC 34), Machinery, except Electrical (SIC 35), Electrical and Electronic Machinery (SIC 36), and Transportation (SIC 37) (see Appendix A).

People interested in an overview of the potential impact that robotics and so-called flexible forms of automation might have on job displacement and productivity in this particular group of industries will find this book useful. Analysts responsible for technology assessment or strategic planning in government, corporations, unions, and "think-tanks," as well as university-based researchers

3

with applied interests in the impacts of automation, are examples of professionals who could benefit from the information presented here. Those who find of interest the types of studies on the impact of technology published by the Organization of Economic Cooperation and Development (OECD) or by the U.S. Congress' Office of Technology Assessment (OTA) would likely find this book of interest.

This is not a book on the technical aspects of robotics and automation. Nor is it a management guide on where, when, and how to implement robotics. People with operational-level responsibilities in a manufacturing company who are in the midst of planning or implementing automated systems will not find in this book the answers or insights they need to complete their projects successfully. Managers and analysts who are looking for detailed answers as to how robot use will affect labor requirements and costs in a particular industry or in a particular company will find the analysis in this book too general. Nonetheless, the practicing manufacturing engineer, production manager, or human-resource planner will most likely find some of the information presented in this volume to be of general interest. For example, it can provide a background and perspective that will help managers answer employees' questions regarding the general impact of robotics on job displacement and on productivity.

Chapter 2 examines the near-term potential impacts of robot use on job displacement and identifies the types of manufacturing jobs that are vulnerable to robotization. In Chapter 2, I review various estimates of the number of related workers that might be displaced by robots over the next 5–15 years and discuss reasons for differences among these estimates. I also summarize the overall findings from the major studies of the impacts of robots on job displacement. The limitations of examining the impacts of one type of technology in particular (e.g., industrial robots) on job displacement are noted, as are the limitations of focusing on displacement induced by technology versus employment changes induced by a variety of economic factors.

Part of the discussion of job-displacement impacts focuses on the cost of using robots. There are substantial economies of scale in the cost of implementing robots within a plant. I argue that only the establishments which can install multiple units for a given application would be able to cost justify them under the conservative view based exclusively on direct labor savings, which is the view of justification that prevailed in the early 1980s and still exists today to a surprisingly large degree. The establishments which can install only one or two units will not be able to cost justify them solely on the basis of direct labor savings. I propose that the scale-economy effect on robot installation costs is useful in explaining the difference between the estimated *theoretical* potential for worker displacement by robots and other estimates of potential displacement, based on forecasts, of the number of robots that may *actually* be used.

Employment data on manufacturing workers are gathered here to highlight the population of people that would be at risk as a result of continuing use of industrial robots. I collected these data on robot acquisition and use directly from users and vendors. Data on the technical potential for job displacement by robots within specific occupational categories come from a survey done in 1981 as part of the Carnegie-Mellon student project course "The Impacts of Robotics on the Workforce and Workplace." Recent data and information on robot costs come from several robot vendors and applications consultants, some of whom are identified in the text and some of whom wished to remain anonymous.

Chapters 3 and 4 examine the impact of so-called flexible forms of automation (e.g., robotics and computer-aided manufacturing systems) on the economics of production. This examination addresses *what would happen if* firms that produce specialized goods in small to medium-sized batches were able to achieve the levels of capital utilization and efficiency currently achieved by firms that produce more standardized goods in higher volumes. A foundation for this analysis is developed in Chapter 3. I have compiled industrywide data from the *Census of Manufactures* (Bureau of the Census, 1981a, b, c, d) and the input-output structure of the United States economy (U.S. Department of Commerce, 1979) and synthesized them into measures of unit cost and levels of output for nearly all of the industries in the metalworking sector. These measures are used to estimate both the current unit cost differential between specialized goods produced in batches and more standardized goods that are mass produced, and the distribution of value added among custom, batch, and mass production in the metalworking industries.

Chapter 4 analyzes in more detail the extent to which production costs could decrease as a result of using the robotic type of automation. The first consideration is cost impacts in cases where robot use results in a moderate decrease in production labor cost as well as a moderate increase in throughput. These cases entail the retrofitting of robots into an existing production system with minimal reorganization of the process. The bulk of the analysis examines the cost impacts in the case where the use of flexible manufacturing systems enables a batch producer of specialized goods to increase levels of machine utilization and output dramatically. This case entails the reorganization of a conventional batch-production factory to take full advantage of robots and other types of computer-aided manufacturing technology, and to utilize its full potential. At the time I carried out this research, it was not possible to make a detailed comparison of the capabilities and economics of flexibly automated plants with those of conventionally organized ones because of a lack of published information due largely to industry inexperience. Thus, my estimate required somewhat unconventional approaches and some stretching of the imagination.

Based on information provided from previous studies, I have estimated the potential for increasing levels of utilization and output in batch production facilities. In Chapter 4, I also expand the analysis of the relationship between unit cost and the level of output across industries in the metalworking sector presented in Chapter 3, and estimate an economies-of-scale parameter. The economies-of-scale parameter shows how unit cost decreases as output levels increase while moving along the spectrum from custom, to batch, to mass modes of organizing production. I combine the scale parameter with the estimated potential for increasing utilization levels in batch-production facilities to arrive at the final estimate, the potential for reducing manufacturing cost. An important part of this analysis is the accumulation of data supporting the claim that potential economic benefits realizable by increased machine utilization are about 10 times the amount of savings that could be realized by reducing direct labor costs.

At the end of Chapter 4, I draw up a plant consolidation scenario. Through this scenario I examine the potential job displacement effects of consolidating the output of several conventionally organized batch-production plants into one high-volume, flexibly automated plant. As a result of flexible automation, the new plant would have the capacity and versatility to handle the output of the several conventionally organized plants, and would employ far fewer workers than were employed in those plants. My chief point is, to the extent that robot use will cause a major problem with worker displacement, it will not be from the substitution of robots for workers in existing plants. Rather, it will likely result from situations where several plants are shut down and their output is consolidated into one plant.

2

Job Displacement Impacts of Industrial Robot Use

2.1. Overview

This chapter begins with a review of the findings of the first group of studies to analyze the impacts of industrial robot use on job displacement. Despite the different approaches used by the various researchers, the message is essentially the same: While the use of robots in the factory will continue to increase steadily, this increase *will not* result in a massive displacement of workers. The key concern relative to worker displacement is that the effects will be highly concentrated in particular occupations, industries, and geographic locations. While there will not be a general problem with worker displacement, displacement will be of sufficient magnitude in some industries and regions and for some groups of workers to be a cause of concern. Overall, relatively few of the workers displaced by robots will actually lose their employment, because most displaced workers will be transferred within the factory to other jobs that have opened up as a result of natural attrition in the work force. Thus, the impact of displacement will primarily be to eliminate new job openings. And while robot use will displace a sizable number of semiskilled and unskilled production workers from their current jobs, the overall impact of robot use on the labor force will be small in comparison with labor-force changes brought about by changes in the general state of the economy.

The bulk of this chapter is a presentation of my analysis of the impacts of robot use on job displacement. That analysis, first presented in Miller, 1983, and substantially revised here, is composed of three major components:

1. a revised estimated percentage of workers that could be displaced by Level I and Level II robots based on the data collected in Carnegie-Mellon University robotics survey in 1981;

7

2. a hypothesis for explaining the difference between the Ayres and Miller (1983) estimated displacement, which is based on the technical potential for using robots, and the Hunt and Hunt (1983) estimated displacement, which is based on a forecast of the size of the robot population;
3. an analysis of the extent to which displacement within a large firm could be offset by attrition and increases in output.

The analysis calls attention to the fact that displacement estimates based on assessments of the proportion of jobs that could be robotized (e.g., the Ayres and Miller approach) are many times larger than displacement estimates based on projections of the actual number of robots that will be in use (e.g., the Hunt and Hunt approach). My argument is that the estimates of potential displacement are not suspect, since the revised estimates of potential displacement for Level I robots derived from the 1981 CMU survey are in close agreement with the 1985 update of the University of Michigan/Society of Manufacturing Engineers (UM/SME) Delphi survey estimates of actual displacement for the year 1995 (Smith and Heytler, 1985). Also, for one important application area—metalcutting machine tool operations—an independently derived estimated potential for robot use is similar to the result derived from the survey data. For these reasons, I argue that the estimates of the potential for Level I robot use are reliable. Therefore, the key to reconciling the difference between these different approaches in estimating displacement levels is to explain why many firms that could potentially use robots will not actually have a strong enough economic incentive to do so.

This chapter also presents an analysis of the cost and benefits of substituting a robot for production workers. The basic approach is to calculate the payback period, based on realistic estimates of the cost of installing robots and on total labor cost saved. I have provided evidence to support the hypothesis that large firms have a substantial cost advantage over small firms in using robots. The underlying reason is that the total cost per robot application decreases as the number of applications increases. Firms that can install multiple robots can do so less expensively on a per-unit basis than firms that install only one or a few units. Since large firms are more likely to have more opportunities for using robots than small firms, they can more likely buy multiple units and therefore realize a lower per-unit investment cost. This analysis provides a rationale to support claims made previously by other analysts (e.g., Hunt and Hunt [1983: 53] and Yoshikawa, Rathmill, and Hatvany [1981: 63]) that for the next several years, robot use will be concentrated primarily within large firms.

If robot use over the next decade were to be concentrated within large establishments (which presumably belong to large firms), the inconsistencies between displacement estimates based on the technical potential for robot use and those based on the forecasted size of the robot population could be reconciled. Almost 40 percent of production workers in SIC 34–37 are employed in large establishments (those with 1000 and more production workers). Across all manufacturing industries (SIC 20–39), only 25 percent of production workers are employed in large establishments. If Level I robots were to be utilized to their full estimated potential, but only within large establishments, this would imply a robot population of 122,000 units if robots were mostly used within industries SIC 34–37 and a population of 213,000 units if robots were used throughout all manufacturing industries. These totals are close to several of the market forecasts of the cumulative robot population for the early to mid-1990s.

The final component of the analysis on displacement impacts examines what would happen to labor requirements in a particular firm if Level I and Level II robots were used to their full potential to perform all the jobs they are capable of doing. For various assumptions regarding the percentage of production workers displaced by robots, the attrition rate, and the impact on the rate of throughput in the factory, the increase in the level of output that would be required to reabsorb workers displaced by robots is calculated. The analysis indicates that Level I robots could be implemented to their full potential within a relatively short period of time (say, 5–10 years) without causing job loss, as long as displaced workers could be reabsorbed into replacement job openings created through attrition. This conclusion would hold even if robot use were to result in a modest increase in the level of output in the factory. In that case, all workers could reasonably be expected to be reabsorbed through the combined effects of attrition and a moderate increase in output stimulated by a price reduction.

2.2 The Eikonix Report, 1979

The report *Technology Assessment: The Impact of Robots* (1979) by the Eikonix Corporation was the first systematic attempt to estimate the work-force impacts of industrial robot use. It identifies the operative worker in manufacturing as "the party most directly affected" by robot use, because people in those occupations "are likely to be the most vulnerable to functional replacement by automation and robotization, based upon the type of work they perform" (Eikonix, 1979: 104, 131). Examples of operative occupations in manufacturing include assemblers, welders and flamecutters, inspectors, packers (excluding produce), metalworking machine operators, sewers, and textile machine operators. The report points out that employment of operatives as a fraction of all employment is highest in the textile and apparel industries and in the metalworking industries: fabricated metals, machinery, excluding electrical and electronic machinery, and transportation equipment.

The Eikonix report gives the first published estimate of the proportion of operative workers that could potentially be displaced by industrial robots (see Table 2-1). The designation of types of robots according to degree of sophistica-

Table 2-1: Eikonix Estimated Percentage of Operative Workers in Manufacturing That Could Be Displaced by Robots

Occupation	Percentage of Candidates for Robotization by:	
	Level I	Level II
Assemblers	0	100
Checkers	0	100
Filers	0	100
Painters	50	50
Drill press operators	100	
Grinder operators	0	100
Lathe operators	50	50
Precision machine operators	50	50
Punch press operators	100	
Solderers	50	50
Stationary linemen	50	50
Welders	50	50
Other operatives	15	55
Machine operatives, miscellaneous specified	0	50
Machine operatives, not specified	0	50
Miscellaneous operatives	0	50
Operatives, not specified	0	50

Source: Eikonix, 1979: Table VIII-7.

tion (or level) was borrowed from a marketing survey which had been published several years earlier (Frost and Sullivan, Inc., 1974). Level I refers to insensate robots—machines that cannot sense the external world. Level II refers to sensor-based systems—machines that can sense the external world, for example, coordinate the mechanical "hand" with a visual "eye." Given a population of about 8 million operatives in manufacturing, the Eikonix report authors estimate that 15 percent of the jobs in manufacturing could be performed by Level I industrial robots, displacing 1.2 million workers, and 55 percent by Level II industrial robots, displacing 4.4 million workers. The authors generated these estimates by examining various job descriptions within each occupational category listed in the U.S. Department of Labor's *Dictionary of Occupational Titles,* as well as from a consideration of the capabilities and limitations of robot technology and presumably from interviews with robot suppliers and users.

The Eikonix report estimates the potential for substituting robots for operative workers in order to estimate the growth of the robot population in the United States. Parameters of the equation used to forecast the robot population were based on the size of the robot population between 1970 and 1976, on assumptions regarding the annual percentage of increase in the size of the population (20 percent), and on the upper limit on the size of the population (400,000 units). This upper limit was determined by dividing the estimate of the number of operatives who could potentially be displaced by Level I robots (1.2 million) by an estimate of the number of workers displaced per robot (3).

According to the Robotic Industries Association (1985), the trade organization of robot producers, there were 14,500 robots in use in the United States as of January 1985. The RIA estimated the population would increase to 20,000 units by the beginning of January 1986. A comparison of the actual robot population, as given by the RIA estimate, with the forecasts made by Eikonix in 1979 is shown below:

	RIA Estimate of Actual Robot Population	1979 Eikonix Forecast of Robot Population		
		Low	Medium	High
January 1985	14,500	10,200	16,200	55,200
January 1986	20,000	12,200	20,300	76,100

The Eikonix projections for the low- and medium-growth scenarios encompass the numbers reported by the RIA. So far, the Eikonix forecast (at least the low- and medium-growth scenarios) has been on target, suggesting that the assumptions underlying the forecast are valid. The key assumption relevant to work-force impacts is that about 15 percent of operative workers in manufacturing can potentially be displaced by Level I robots by the year 2000.

The Eikonix high-growth scenario forecast is about four times greater than the current robot population. Does this large variance indicate that their underlying assumptions used for the forecasts are invalid? The report's authors note that managers will not necessarily want to replace workers with robots even where it is feasible to do so. Substituting robots for workers also substitutes a fixed cost in the form of a capital asset for an operating cost which presumably resembles a rental cost that can be turned on and off at will. This substitution of fixed for variable costs "may be a strong argument against robotization" (Eikonix, 1979: 135). The report also notes that other forces, such as competition with other types of machines, difficulties with obtaining investment capital and with integrating robots into existing organizations, labor and management resistance, and inadequate technical information, will also constrain the growth of robot use. Presumably, the high-growth scenario, assumes that these factors would all be resolved in a way that does not impede the growth of robot use (which has not been the case). The differences between the actual robot population and the projections of the high-growth scenario highlight the dangers of focusing solely on the technical potential for substitution without giving explicit consideration to institutional, organizational, and behavioral factors.

The conclusion of the Eikonix report is that the fear of mass displacement of workers by robots is completely inappropriate within the 20-year time frame considered and perhaps for some years beyond. The authors comment:

Best estimates of robotic deployment are for a slow, but steady diffusion into the industrial picture with a much slower diffusion into the service industries. Local or structural employment effects will be noted, but over a period of time these will surely be swamped in phenomena which are more closely related to the state of the economy. Robots will not cause massive numbers of people to lose their jobs. (Eikonix, 1979: 216)

2.3 The Carnegie-Mellon and Ayres and Miller Studies, 1981–1983

. The next major study of robot use and job displacement to have taken place was "The Impacts of Robotics on the Work-force and Workplace," released by Carnegie-Mellon University (1981).[1] Part of the report focuses on work-force impacts in order to provide "knowledge of where there are large numbers of workers who do the types of jobs that robots can do . . ." (p. 84). Like the Eikonix study, the CMU study focuses on operative jobs in manufacturing and provides an estimate of the percentage of jobs within major operative occupational categories that could be done by Level I and Level II robots. These percentage estimates were used in conjunction with employment data to determine an upper limit on the number of jobs which could potentially be displaced by robot use. Unlike the Eikonix study, the CMU study derived these estimates from information gathered from experienced and prospective robot users. Also, the CMU study gives more attention to where these potentially vulnerable jobs are located by industry and by geographic region.

Estimates of the percentage of jobs within specific occupational categories that could be performed by robots are shown in Table 2-2. All respondents were either already using robots or in the process of evaluating robot applications. A key point is that the respondents were asked to estimate the percentage of jobs within specified occupations that *could* be robotized, acknowledging that robots would not be used in every one of the positions. The CMU report (1981: 95) notes:

> In follow-up interviews, some respondents said they thought it was difficult and somewhat unrealistic to make the distinction between the jobs where robots could be used, and the jobs where it is economically justifiable to use them. We cannot really say whether these estimates are an upper bound on *perceived* robot applicability, or an estimate of the percentage of jobs where it could be economically justifiable to use them.

To estimate the number of jobs that could be displaced, the estimates shown in Table 2-2 were multiplied by the number of people in each occupational category. This was done in two steps. The first step was to restrict the focus to

---

1. The report is the result of a project staffed and managed by CMU students, under the guidance of several CMU professors.

Table 2-2: Carnegie-Mellon Estimated Percentage of Selected Manufacturing Production Workers' Jobs That Could Be Performed by Robots

Occupation	Level I Robots		Level II Robots	
	Range of Responses	Average Weighted Response	Range of Responses	Average Weighted Response
Inspector	5–25	13	5–60	35
Tester	1–10	8	5–30	12
Assembler	3–20	10	20–50	30
Caster	5–15	5	10–20	10
Coiler/winder	15–40	24	15–50	40
Conveyor operator	10–50	16	20–65	25
Packager	1–40	16	2–70	41
Order filler	5–20	12	5–80	32
Machine tool operator (non–NC)	10–30	15	5–60	30
Machine tool operator (NC)	10–90	20	30–90	49
Welder/flame-cutter	10–60	27	10–90	49
Filer/grinder/buffer	5–35	20	5–75	35
Production painter	30–100	44	50–100	66
Mixer	5–10	10	5–10	10
Sandblaster/shotblaster	10–100	35	10–100	35
Millwright	0–15	1	0–15	1
Oiler	—	0	—	0
Rigger	—	0	—	0
Pourer	5–20	10	10–30	24
Kiln/furnace operator	0–10	2	0–20	15
Tool and die maker	1–5	3	4–60	50
Electro-plater	5–40	20	5–60	55
Dip plater	20–100	40	50–100	77
Etcher/engraver	15–100	29	15–100	90
Heat treater/annealer	5–50	10	5–90	46
Drill press operator	25–50	30	60–75	65
Grinding/abrading/cutting machine operator	10–20	18	20–100	50
Lathe/turning machine operator	10–20	18	40–60	50
Mill/plane machine operator	10–20	18	40–60	50
Riveter	5–100	15	100	30
Punch press operator	10–100	40	60–80	70
Electronic wirer	0–10	9	10–50	28

Source: Carnegie-Mellon University, 1981: Table 5-20.
Based on a total of 16 respondents.
Not all respondents gave estimates for all occupations.

employment in the metalworking industries (SIC 34–37),[2] since at the time of the study nearly all robots in the United States and worldwide were located in these industries. The result of this step is shown in Table 2-3. This table shows the number of people employed in nine operative occupational categories that are identified from the survey responses as being "prime candidates for robotization," along with the estimated number of workers that could be displaced by robots. The estimated potential displacement for each occupational category was derived simply by multiplying the number of people employed in that category by the average survey response for the percentage of jobs that could be robotized. The data are given for states which have the largest number of production workers in the metalworking industries, and for the whole nation as well. The result is that Level I robots could potentially displace about a half million people working in these industries, based on 1977 employment data, and that a large fraction of these workers are geographically concentrated in the five Great Lakes states (Indiana, Illinois, Michigan, Ohio, and Wisconsin), plus New York and California.

The second step was to do a similar calculation using employment figures for operatives in all manufacturing industries as opposed to just operatives in metalworking industries. This calculation is shown in Table 2-4. The resulting estimate is that about a million manufacturing workers are potentially vulnerable to displacement by commercially available Level I robots, and that about 3 million workers are potentially vulnerable to displacement by the next generation of sensor-based Level II robots. The report's overall conclusions on the work-force impacts of industrial robot use are as follows (Carnegie-Mellon University, 1981: 17):

Nearly 7 percent of the total work force do the types of jobs which currently are, or soon will be in the domain of industrial robots. But robots cannot do all of these tasks in the foreseeable future, especially if they are retrofitted into existing factories. More realistically, robots which are commercially available today could possibly perform nearly 16 percent of the operative tasks within those manufacturing industries where they are well suited (with estimates ranging between 8 and 32 percent), and sensor based robots could perform nearly 40 percent of these operative tasks (with estimates ranging from 20 to 80 percent). In the short term, maybe as many as two percent of the entire work force could possibly be replaced by robots. Within the next two decades, maybe the number will increase to 4 percent, possibly to 7 percent. This is hardly catastrophic on

2. This includes Fabricated Metals (SIC 34), Machinery, except Electrical (SIC 35), Electrical and Electronic Equipment (SIC 36), and Transportation Equipment (SIC 37). The numbers following each industry are its standard industrial classification number designation. The major group Measuring, Analyzing, and Controlling Instruments (SIC 38) is also regarded as part of the metalworking industries. However, since relatively few people are employed in SIC 38, most of the subsequent discussion of employment impacts focuses on industries in SIC 34-37. In the several instances where a reference to the metalworking industries does include SIC 38, it is so noted.

Table 2-3: Carnegie-Mellon Estimated Number of Production Workers in Industries SIC 34–37 That Could Be Displaced by Robots

	Employment in 3 Categories	Number of Workers[a] That Could Be Displaced by:		Employment in 6 Categories	Number of Workers[b] That Could Be Displaced by:		Employment in All 9 Categories[c]	Number of Workers[c] That Could Be Displaced by:	
		Level I	Level II		Level I	Level II		Level I	Level II
New York	62,680	13,997	31,859	85,610	9,880	27,402	148,290	23,877	59,261
New Jersey	33,590	7,372	17,046	42,130	4,842	13,485	75,720	12,214	30,531
Pennsylvania	12,240	2,687	6,168	4,440	581	1,497	16,680	3,268	7,665
Georgia	17,560	4,113	8,977	21,560	2,399	6,760	39,120	6,512	15,737
North Carolina	20,910	4,633	10,602	30,870	3,373	9,599	51,780	8,006	20,201
Tennessee	11,120	2,568	5,704	25,430	2,859	8,027	36,550	5,427	13,731
Indiana	61,078	13,515	30,982	84,744	9,579	26,948	145,822	23,094	57,930
Illinois	99,261	22,133	50,387	100,379	11,722	32,526	199,640	33,855	82,913
Michigan	144,100	32,023	73,072	104,750	16,493	44,794	248,850	48,516	117,866
Ohio[d]									
Wisconsin	56,490	12,806	28,645	49,365	5,658	15,766	105,855	18,464	44,411
Texas	16,620	3,951	8,508	16,520	1,894	5,209	33,140	5,845	13,717
Missouri	20,910	4,993	10,728	38,560	4,360	12,188	59,470	9,353	22,916
California	95,188	21,854	48,595	148,749	17,170	47,187	243,937	39,024	95,782
13-state total	651,747	146,645	331,273	753,107	90,810	251,388	1,404,854	237,455	582,661
U.S. Metal-working total	1,165,630	262,261	573,972	1,582,010	182,873	504,506	2,747,640	445,134	1,078,478

Source: Carnegie-Mellon University, 1981: Table 5-26.
aOccupations include: production painters, welders, flamecutters, machinists, machine tool operators (combination and numerical control and tool room), drill press operators, grinding/abrading machine operators, lathe/turning machine operators, milling/planing machine operators, punch press operators.
bOccupations include: electroplaters, heat treaters, buffers, packagers, filers/grinders/buffers, inspectors, assemblers.
cSum of other two sets of figures.
dInformation not available.

Table 2-4: Carnegie-Mellon Estimated Number of Production Workers in All Manufacturing Industries That Could Be Displaced by Robots

Occupation	Total Employed, 1979[a]	Percentage of Workers That Could Be Displaced by[b]:		Number of Workers That Could Be Displaced by:	
		Level I	Level II	Level I	Level II
Assemblers	1,289,000	10	30	128,900	386,700
Checkers, examiners, and inspectors[c]	746,000	15	35	111,900	261,100
Packers and wrappers (except meat and produce)	626,000	15	40	93,900	250,400
Production painters	185,000	45	65	83,250	120,250
Welders and flamecutters	713,000	25	50	178,250	356,500
Machinists and machine operators[d]	3,027,000	20	50	605,000	1,513,500
	6,586,000			1,201,600	2,888,450

Source: Carnegie-Mellon University, 1981: Table 5-27.

[a]Employment figures from Statistical Abstract of the United States, 1980 (Bureau of the Census, U.S. Department of Commerce, Washington, DC: Government Printing Office, 1980), Table 697, "Persons Employed by Sex, Race, and Occupation, 1972 and 1979."

[b]Based on survey results within the metalworking sector. Percentage estimates are the average weighted responses of survey results, rounded to nearest multiple of 5.

[c]Also includes testers. Since there are more than four times as many inspectors as testers in metalworking, the percentage estimate for inspectors is used.

[d]Includes the following occupational categories: machinists and jobsetters, precision machine operators, machine operators, and punch press operators.

the national scale, especially when one allows for new job opportunities. But adjustments made in response to rapid diffusion of this new technology are likely to be intensified in those geographic regions, such as the Great Lakes States, the northeastern United States, and California, where manufacturing operations are concentrated.

The message of the analysis is twofold. First, even considering upper limits of what is technically feasible (as opposed to a lower limit of what is economically viable), only a small fraction of the work force do the types of jobs that robots can presently do, or will soon be able to do. This is the same message as in the Eikonix report. Second, to the extent that displacement would be a problem, it would be so because the vulnerable occupations are concentrated within a subset of industries which are, in turn, geographically concentrated in several regions of the country.

An interesting point is the difference between Ayres and Miller interpretation of results of the project report and that of the popular media. In an article popularizing the results (Ayres and Miller, 1982), we state that the "transition of substituting robots for semiskilled manufacturing jobs will not be catastrophic if workers are trained and directed toward growth areas. In fact, this transition will probably be less dramatic than the impacts of office automation." Ayres and I have commented that in the time horizon required for sophisticated robots to displace a large fraction of manufacturing operatives (40 years or more), most current operatives will have retired or left their jobs, and robot-related jobs and new growth sectors in the economy will provide new job opportunities. In contrast, citing the conclusions in the 1981 CMU report, *Newsweek* (1981: 90) reported, "Such a reduction would mean massive and permanent layoffs for the semiskilled and unskilled workers who now ply the nation's production lines. . . ."

Ayres and I have expanded the discussion of employment impacts of robotization in our book *Robotics: Applications and Social Implications* (1983). Displacement estimates there were derived in the same manner as in the previous CMU project report. One difference in the 1983 work is that we extrapolated the survey estimates of potential displacement to include occupations that were not part of the original survey "based on the original responses, and on our own best judgements of occupational similarities" (Ayres and Miller, 1983: 203). Additional occupations in the semiskilled (operative), skilled (metalworking craftworker), and unskilled (laborer) categories were included as part of the vulnerable-worker population for robotic substitution. Thus, the estimates for potential substitution from the CMU survey (14 percent for Level I and 40 percent for Level II) were multiplied by a somewhat larger population of workers; we argue that Level I robots could potentially do the work of 1.5 million manufacturing workers, and Level II robots could potentially displace 4 million.

In the updated work, we also consider government estimates of economywide job growth in the industries and occupations where robots will be applied. Citing forecasts of the Department of Labor's Bureau of Labor Statistics (BLS), we note that the same industries most likely to be subject to robotization in the 1980s are also those in which above-average increases in employment are expected. We compare their potential displacement estimates with forecasted increases in employment for blue-collar occupations (shown in Table 2-5). The comparison shows that *even if* all of the potential for Level I displacements were to be realized in a short period of time (by 1990), the number of workers displaced would be nearly offset by the forecasted increases in employment. We also note that there is a high rate of annual labor turnover within several of the semiskilled and unskilled types of jobs that could be robotized.

The data presented in the 1983 study reinforce the idea set forth earlier that work-force impacts of robots are not the result of the absolute number of people affected, but of the "concentration of effects in a relatively narrow industrial sector, occupational, and regional setting" (Ayres and Miller, 1983: 188). Because of the concentration of effects, we argue that "displacement will be of sufficient magnitude, at least in some industries and regions, and for some groups of workers, to be a cause of concern." A similar conclusion regarding the potential impacts of robots on job displacement has been reached by Volk-

Table 2-5: Ayres and Miller Comparison of Potential for Employment Growth and Robotic Displacement by 1990

	Employment, 1980 (in thousands)	Incremental Increases in Employment Projected for 1990 (in thousands)[a]		
		Low	High II	High I
Metalworking: SIC 33–38				
Operatives[b] and laborers	4,673.6	785.5	901.5	1,373.1
Craftworkers[c]	2,015.2	359.4	408.0	608.9
Blue-collar, total	6,688.8	1,144.9	1,309.5	1,982.0
Estimated number potentially displaced by:				
Level I robots	800.0			
Level II robots	2,000.0			
Manufacturing, Total: SIC 20–39				
Operatives[b] and laborers	10,421.9	1,296.5	1.447.9	2,234.7
Craftworkers[c]	3,768.4	536.1	605.6	888.1
Blue-collar, total	14,190.3	1,832.6	2,053.5	3,122.8
Estimated number potentially displaced by:				
Level I robots	1,500.0			
Level II robots	4,000.0			

Source: Ayres and Miller, 1983: Table 5-11. Compiled from Personick, 1981, and from *The National OES Based Industry-Occupation Matrix for 1980* (Bureau of labor Statistics, U.S. Department of Labor, Washington, DC: Government Printing Office, 1982).
[a]Assumptions for low, high I, and high II growth scenarios are given in Personick, 1981.
[b]Operatives include transport operatives.
[c]Craftworkers include both metalworking craftworkers and other craftworkers.

holz (1982), based on a study of robot use in West Germany. He states: "From the overall social standpoint, and in the light of this time scale, the figures [of potential displacement] no longer seem so catastrophic. On the other hand, it has to be remembered that the use of robots will be concentrated in a small number of industries" (p. 188).

2.4. The Upjohn Institute Study, 1983

The W. E. Upjohn Institute for Employment Research (Kalamazoo, Michigan) prepared two studies which specifically estimate the job displacement and job creation potential of industrial robots in the United States by 1990. The first study presents the results for the state of Michigan (Hunt and Hunt, 1982). The second study, an extension of the first, presents the results for the entire United States (Hunt and Hunt, 1983). The studies were initiated and funded by the state of Michigan. The state's Occupational Information Coordinating Committee believed that "robotic technology might significantly affect the state's economy" and therefore commissioned the Upjohn Institute "to look at the labor market implications of robotics in order to provide a base upon which human resource planning could proceed" (Hunt and Hunt, 1983: 5). The discussion of the work that follows was drawn from the 1983 publication, since this is the more comprehensive analysis. Also, this review focuses on their analysis of job displacement.

In the Eikonix study and Ayres and Miller studies discussed above, attempts were made to estimate an upper limit on the number of production workers in manufacturing that could potentially be displaced by industrial robots. In the studies of both Eikonix and Ayres and Miller, this estimate was made by considering the fraction of jobs within selected occupational categories that could be robotized independent of considerations of the number of robots that would actually be in use. In the Eikonix study, estimates of potential displacement were used to forecast the growth in size of the robot population.

Hunt and Hunt (1983) take an opposing approach in their estimates of job displacement. They explain:

Unlike the Carnegie-Mellon study, the projections of occupational displacement in this study are the result of first forecasting the U.S. robot population by industry and application areas within these industries. This approach constrains the displacement estimates to reflect the actual expected sales of robots. In this way, a consistent economic framework is established within which it is possible to estimate not only the population of robots and job displacement, but also the job creation resulting therefrom. (p. 29)

First, Hunt and Hunt estimate an upper and lower bound on the total number of robots in use in the United States by 1990. This is a "bottom up" estimate in the sense that it is derived from low and high estimates of the number of robots that will be in use in five categories of applications and in two types of industries (automotive and all other manufacturing). Their forecast of the U.S. robot population in 1990 is shown in Table 2-6. According to their forecast, by 1990 the total robot population in the United States will range from a minimum of 50,000 units (low-growth scenario) to a maximum of 100,000 units (high-growth scenario).[3] They assume that new breakthroughs in robotic technology will not invalidate the robot population forecasts, because they believe that "diffusion of robotic technology will be limited more by a lack of human understanding of *existing* technology than by a lack of new hardware" (Hunt and Hunt, 1983: 46).

3. The low-growth scenario assumes relatively high interest rates, slow real GNP growth, approximating the late 1970s annual average of 2.0 percent, and lagging auto sales. The high-growth scenario assumes declining interest rates and faster real GNP growth, approximating the post-World War II annual average of 3.5 percent.

Table 2-6: Hunt and Hunt Forecast of the U.S. Cumulative Robot Population in 1990

Application	Autos Range of Estimate		All Other Manufacturing Range of Estimate		Total Range of Estimate	
	Low	High	Low	High	Low	High
Welding	3,200 (21.3%)	4,100 (16.4%)	5,500 (15.7%)	10,000 (13.3%)	8,700 (17.4%)	14,100 (14.1%)
Assembly	4,200 (28.0%)	8,800 (35.2%)	5,000 (14.3%)	15,000 (20.0%)	9,200 (18.4%)	23,800 (23.8%)
Painting	1,800 (12.0%)	2,500 (10.0%)	3,200 (9.1%)	5,500 (7.3%)	5,000 (10.0%)	8,000 (8.0%)
Machine loading/unloading	5,000 (33.3%)	8,000 (32.0%)	17,500 (50.0%)	34,000 (46.0%)	22,500 (45.0%)	42,000 (42.0%)
Other	800 (5.3%)	1,600 (6.4%)	3,800 (10.9%)	10,500 (14.0%)	4,600 (9.2%)	12,100 (12.1%)
Total	15,000	25,000	35,000	75,000	50,000	100,000

Source: Hunt and Hunt, 1983: Table 2-10.

Hunt and Hunt's forecast of the size of the U.S. robot population in 1990 is compared with some of the forecasts made in Table 2-7. Their forecast is conservative when compared with some of the forecasts made in the 1980–82 time frame (e.g., the 1982 UM/SME Delphi survey), which predicts 150,000 units by 1990. The Hunt and Hunt forecast now seems "middle-of-the-road" when compared with the more recent Tech Tran Corporation (1983) estimate. It is slightly more conservative than the medium- and high-growth forecast of Leontief and Duchin (1984), and could be viewed as "optimistic" given the results of the revised 1985 UM/SME Delphi survey. The important point is that the Hunt and Hunt forecast is comparable to other recent forecasts.

Hunt and Hunt claim that actual robot use in the United States will steadily grow, but, for six reasons, perhaps less spectacularly than has been anticipated by others:

Table 2-7: Comparison of the Hunt and Hunt Forecast with Other Forecasts of the 1990 Robot Population

	Low	Medium	High
Eikonix (1979)[a]	24,500	48,000	209,000
Robot Institute of America (1981)[b]	75,000	—	100,000
UM/SME Delphi (1982)[c]	—	—	150,000
Hunt and Hunt (1983)[b]	50,000	—	100,000
Tech Tran (1983)[c]	—	—	69,000
Leontief and Duchin (1984)[d]	22,000	69,000	127,000
UM/SME Delphi (1985)[e]	30,000	40,000	60,000

[a]Eikonix, 1979: 236–238

[b]As reported in Hunt and Hunt, 1983: 34.

[c]Tech Tran Corporation, 1983: 21; only a point estimate is given.

[d]Leontief and Duchin, 1984: 1.23.

[e]Smith and Heytler, 1985: 167. The medium value is the median response. The low and high values are the interquartile ranges (the middle 50 percent of the response distribution).

1. the lack of trained personnel both to implement robotic technology and to maintain and support that technology once installed;
2. the large financial commitment necessary to implement robotics;
3. the extensive management commitment needed to adopt robots successfully;
4. general economic conditions, which will result in incremental and gradual investment;
5. the unlikelihood of a much more rapid diffusion of robot technology than in earlier process technologies;
6. the limited diffusion of robotic technology to large firms, perhaps just Fortune 500 firms, for the foreseeable future.

Given the forecast of the robot population, Hunt and Hunt estimate displacement in two additional steps. First, they estimate the average job displacement effect of each robot. Based on interviews with robot users, they conclude that for the horizon of the forecast one robot will displace one worker per shift and that most industries will operate two shifts per day. The estimate of the number of workers displaced is obtained simply by multiplying the figures in Table 2-6 by a factor of two. Second, they estimate the relative magnitude of job displacement within each of the five occupational categories and within each industry group by comparing displacement totals with employment totals. Their results are shown in Table 2-8.

Finally, Hunt and Hunt compare displacement rates in their selected occupation and industry groups with Bureau of Labor Statistics estimates of the rate of job openings. These results are shown in Table 2-9. The displacement rates are adjusted to a simple average annual job displacement rate by dividing the esti-

Table 2-8: Hunt and Hunt Estimated Percentage of Manufacturing Production Workers Displaced by Robots by 1990

Occupation	Autos		All Other Manufacturing		Total	
	1980 Employment Level	Displacement Range (%)	1980 Employment Level	Displacement Range (%)	1980 Employment Level	Displacement Range (%)
Welding	41,159	15–20	359,470	3–6	400,629	4–7
Assembly	175,922	5–10	1,485,228	1–2	1,661,150	1–3
Painting	13,556	27–37	92,622	7–12	106,178	9–15
Machine loading/ unloading	80,725	12–20	988,815	3–7	1,069,540	4–8
All operatives and laborers	467,846	6–11	9,954,048	1–2	10,421,894	1–2
All employment	773,797	4–6	19,587,771	0–1	20,361,568	0–1

Source: Hunt and Hunt, 1983: Table 3-3. Compiled from employment data based upon unpublished OES data provided by Office of Economic Growth and Employment Projections, Bureau of Labor Statistics, U.S. Department of Labor, Washington, D.C.

Table 2-9: Hunt and Hunt Comparison of Annual Rate of Worker Displacement with Replacement and Growth Needs for Selected Manufacturing Occupations

	Simple Average Annual Displacement Impact of Robots 1980–1990			BLS Average Annual Replacement Needs 1978–1990	BLS Total Average Annual Openings 1978–1990
Occupation	Autos	All Other Manufacturing	Total	All Industries	All Industries
Welding	2.0	.6	.7	2.3	5.1
Assembly	1.0	.2	.3	3.0	6.5
Painting	3.7	1.2	1.5	2.4	3.9
Machine loading/ unloading	2.0	.7	.8	2.5	3.0
All operatives and laborers	1.1	.2	.2	2.9	4.0
All employment	.7	.1	.1	3.8	5.5

Source: Hunt and Hunt, 1983: Table 3-8. Based on replacement needs and total average annual openings from *The National Industry-Occupation Employment Matrix, 1970–1978, and Projected 1990*, U.S. Department of Labor, Bureau of Labor Statistics, Bulletin 2086, Vol. 2, April 1981, pp. 495–502.

mates in Table 2-8 by 10, because these figures represent total displacement over a 10-year period. The table shows the BLS estimate for average annual replacement needs, which is the percentage of current employment needed to replace those who leave the work force as a result of death, disability and retirement. The BLS estimates for total rates of job openings (replacement plus projected changes in demand) are also shown. The table shows that in no case do the Hunt and Hunt displacement estimates exceed the BLS projections for job openings, and, except for painters, replacement needs alone are larger than projected displacement. The conclusion is that the impact of displacement will primarily be to eliminate job openings.

Hunt and Hunt compare their estimated displacement rates with other estimated displacement rates published in the 1981 Carnegie-Mellon study and in the 1982 UM/SME Delphi survey (Smith and Wilson, 1982). This comparison is shown in Table 2-10. The Hunt and Hunt estimates were derived from their forecast of the number of robots in use. The Carnegie-Mellon study estimates were derived from a survey asking current and prospective robot users about the percentage of jobs that could be done by robots. The UM/SME estimates of actual and potential displacement were derived from a Delphi survey of current and prospective robot users, robot producers, and other experts.

The CMU survey estimates of potential displacement are rather close to the Hunt and Hunt estimates for the automobile industry. Hunt and Hunt (1983: 84–85), explain:

This result may be due in part to the Carnegie-Mellon weighting procedure, which gives greater importance to large employers. The large firms in their survey are more

Table 2-10: Comparison of Hunt and Hunt with Other Surveys' Estimated Percentages of Worker Displacement

Occupation	W. E. Upjohn Institute Estimates		CMU Survey (Level I)[a]	UM/SME Delphi Forecast[b]	
	Auto	Total		Potential	Actual
Welding	15–20	4–7	27	20	10
Assembly	5–10	1–3	10	10	5
Painting	27–37	9–15	44	20	15
Machine loading/unloading	12–20	4–8	20	10	6

Source: Hunt and Hunt, 1983: Table 3-4.
[a]The displacement rates shown are those from the weighted average response for Level I robots. See Carnegie-Mellon University, 1981: 97–99.
[b]From Smith and Wilson, 1982: 70.

likely to be auto or auto-related firms, since there is a disproportionate concentration of both robot users and large establishments within the auto industry. This raises the possibility that the Carnegie-Mellon displacement estimates could be more descriptive of the auto industry than an all-industry average, at least through 1990.

. . . Given that the auto industry is the leader in the application of robots, this may corroborate their theoretically possible levels of displacement for Level I robots.

A detailed breakdown of the respondents to the 1981 CMU survey who provided displacement estimates, given in Miller, (1983: p. 235), shows that the majority of respondents were, in fact, from large establishments. However, the majority of respondents from large establishments were not from the automobile industry (Appendix B).

The displacement estimates from the 1982 UM/SME Delphi survey were based on responses from a cross section of firms in the metalworking industries (SIC 34–37). The sample size in the Delphi survey was several times larger than that of the 1981 CMU survey.[4] Respondents to the Delphi survey were asked two questions relating to worker displacement. First, they were asked to forecast the extent to which it would be *economically possible or feasible* to displace workers with robots by the years 1990 and 1995 within six selected occupations. This is what is labeled as the "potential" displacement rate in Table 2-10. Second, they were asked to forecast the extent to which workers within the selected occupations *actually will be displaced or transferred* by the years 1990 and 1995 as a result of robot use. This is labeled as the "actual" displacement rate in the table. The Delphi estimates of potential and actual displacement shown in Table 2-10 are the median estimates for the year 1990.

4. The Delphi survey consisted of three rounds, and Smith and Wilson (1982: 1) note that the number of participants in each survey round ranged from 36 to 60. Thus, there were at least twice as many responses as in the survey data used by Ayres and Miller, possibly four times as many (Carnegie-Mellon University, 1981; Ayres and Miller, 1983). Eighty to 90 percent of the participants in each round of the Delphi survey were from robot-using firms.

One important point is the difference between the estimates of potential (economically justifiable or feasible) displacement versus actual displacement in the Delphi survey. For example, in the two occupational areas with the largest number of production workers—assembly and machine operations—actual displacement estimates are respectively, 50 and 40 percent less than the potential displacement estimates. Respondents clearly indicated that many robot applications which will be feasible and justifiable in 1990 will not be implemented. Respondents also indicated that a sizable differential will exist between the potential for robot use and actual robot use in 1995, although the differential is smaller than anticipated for 1990.

Another interesting point is the difference in the magnitude of the Delphi estimated displacement for the years 1990 and 1995. Table 2-11 compares the 1981 CMU survey results with the 1982 Delphi estimated potential and actual displacement for the year 1995. The CMU estimates for potential Level I robot use are very close to the Delphi survey estimates for potential displacement in the year 1995 for three of the four occupations shown in the table, with painting being the exception.

Hunt and Hunt (1983) have two major objections to the estimated potential displacement first reported in the 1981 CMU study, and extended by Ayres and Miller (1983) in subsequent work. The first objection is that ". . . the emphasis is clearly on theoretical displacement in the indefinite future rather than actual or probable displacement by some specific date." The discussion of the Delphi survey results highlights the difference between asking respondents for their opinions on the number of workers that *could* be displaced by robots versus the number of workers that *actually will* be displaced. Apparently, institutional, organizational, and behavioral factors cited by the Eikonix (1979) and the Hunt and Hunt (1983) studies account for this differential. Only time will tell if the estimates of potential displacement prove to be practically unobtainable goals or if the estimates of actual displacement prove to be overly conservative. The difference in results obtained when asking respondents to consider displacement by a specific date in the near term (e.g., 1990) versus a time further into the future (e.g., 1995) is also highlighted in the discussion of the Delphi survey results. Given these factors, the difference between the results

Table 2-11: Comparison of Estimated Potential Worker Displacement from 1981 CMU Survey and 1982 UM/SME Delphi Survey

Occupation	CMU Survey (Level I)	UM/SME Delphi Forecast for Year 1995	
		Potential	Actual
Welding	27	25	15
Assembly	10	15	12
Painting	44	25	20
Machine loading/unloading	20	20	10

obtained by Hunt and Hunt and by Ayres and Miller is not at all surprising.

Hunt and Hunt's second objection is that they do not feel that the Ayres and Miller displacement estimates can be generalized across all manufacturing industries. They comment (1983: 26), "We doubt that production techniques, even theoretically, are as homogeneous across manufacturing as Ayres and Miller imply; by industry, by size of firm, or by type of product." The similarity between the Ayres and Miller estimates for Level I robot displacement and the Delphi survey estimates for potential displacement for 1995 provides some support for the claim that both surveys are reliable indicators of the perceived potential for robot use in the metalworking industries (SIC 34–37). What is objectionable, according to Hunt and Hunt, is Ayres and Miller's use of the displacement estimates, based on responses from metalworking firms to estimate potential displacement across all manufacturing industries (SIC 20–39). Robots are thought to be primarily applicable in the discrete production of durable goods, and most of this type of production takes place within the industries in SIC 34–37. If the survey data are used, the estimated potential displacement in industries outside of SIC 34–37 can, therefore, be expected to given an upper bound on potential displacement, since the applicability of robots in these industries is presumably less than in those industries in SIC 34–37.

Interestingly enough, despite the major differences in perspectives and methodology, the conclusions on job displacement reached by Hunt and Hunt are essentially the same as those reached by Ayres and Miller. Hunt and Hunt (1983: 174) comment:

. . . While we are convinced there will be no general worker displacement problem, there clearly *will* be particular pockets of displacement that may cause labor market distress. Particular occupations, industries, and locations will suffer the brunt of job displacement impact. Examples include industrial welders and production painters, the auto industry, and Southeast Michigan. In each of these cases, substantial job displacement will occur in the decade of the '80's because of the application of robots. While a review of labor force attrition rates suggests that there will be very few workers actually thrown out of work even in these highly impacted areas, there is still some potential for displaced workers in these situations.

Hunt and Hunt were the first researchers to systematically estimate the type and number of jobs likely to be created as a result of producing and using robots. Their broader conclusions on the impacts of robot use on the work force were reached after comparing the results of the analysis of job displacement with job creation. They state (1983: 172, 176):

The most remarkable thing about the job displacement and job creation impacts of industrial robots is not the fact that more jobs are eliminated than created; this follows from the fact that robots are labor-saving technology designed to raise productivity and lower costs of production. Rather, it is the skill-twist that emerges so clearly when the jobs eliminated are compared to the jobs created. The jobs eliminated are semi-skilled or unskilled, while the jobs created require significant technical background. We submit that this is the true meaning of the so-called robotic revolution.

2.5. The Miller Study, 1983

My study "Potential Impacts of Robotics on Manufacturing Costs Within the Metalworking Industries" (Miller, 1983) was the next major analysis of robot use and job displacement. The analysis of impacts on work-force requirements consisted of three major components:

1. a revised estimate of the percentage of workers that could be displaced by Level I and Level II robots based on the data collected from the CMU robotics survey in 1981;
2. a hypothesis for explaining the difference between the Ayres and Miller (1983) estimated displacement based on the technical potential for using robots and the Hunt and Hunt (1983) estimated displacement based on a forecast of the size of the robot population;
3. an analysis of the extent to which displacement within a large company could be offset by attrition and increases in output.

I have reviewed these three components of the analysis in detail here. While some of the details I have presented differ from the 1983 study and subsequent publications based on it (e.g., Miller, 1985), the basic conclusions reached in the 1983 study are still the same.[5]

2.6. Revised Estimated Potential Displacement

The estimated potential displacement within the metalworking industries (SIC 34–37) by occupation presented in previous Ayres and Miller studies was revised in Miller, 1983 for several reasons. First, additional survey responses are included, increasing the total number of responses from 16 to 24. Details on the survey data, including background information on the firms that responded, are given in Appendix B. Second, a more systematic job was done of extrapolating the results for occupations included in the survey to other metalworking

5. Some of the numbers used in the 1983 study have been revised here. For example, in the original study, some calculations requiring industry-level employment data were made using figures from a particular industry, engines and turbines (SIC 351). These calculations have been redone here using employment totals from all the metalworking industries considered in this analysis (SIC 34–37). It is more appropriate to use employment totals for all industries in SIC 34–37 than for a particular industry, since respondents to the survey were spread across all industries in SIC 34–37.

Table 2-12: 1980 Production Worker Employment in Industries SIC 34–37

Occupation	Number of Production Workers Employed	Percentage of Total Employment
Tool handlers	383,490	7.5
Metalcutting machine operators	1,073,380	21.0
Metalforming machine operators	337,841	6.6
Other machine operators	785,660	15.4
Assemblers	974,410	19.1
Inspectors	216,800	4.2
Material handlers and laborers	479,960	9.4
Miscellaneous craftworkers	507,980	10.0
Maintenance and transport workers	341,740	6.7
Total	5,101,261	100.0a

aDoes not add to this total because of rounding.

occupations.[6] Third, in the previous studies, employment data from the year 1977 were used. In the revision, I used 1980 employment data, which correspond more closely to the time the survey data were collected.

All occupational titles for production workers in industries SIC 34–37 given in the 1980 *Occupational Employment* survey (Bureau of Labor Statistics, 1982b) are grouped into nine major categories and shown in Table 2-12. (Details on which occupational titles are assigned to each major category are given in Appendix C.) The table shows that, as of 1980, metalcutting machine operators and assemblers composed the two largest groups of production workers in these industries. The results of my revised estimated potential displacement within each of the nine major occupational categories are shown in Table 2-13. The table also shows the number of people in SIC 34–37 employed in each of these categories as of 1980. The figures for potential displacement shown in the last two columns were obtained by multiplying the estimated percentages of displacement by the employment totals. The result is that Level I robots could potentially do the jobs of about 12 percent of production workers, displacing over 625,000 workers in the metalworking industries (SIC 34–37). Likewise, Level II robots could potentially do the jobs of about 33 percent of the production workers, displacing about 1.7 million workers in these same industries.

The extrapolation of the results to production workers in all manufacturing is shown in Table 2-14. These results were obtained by assuming that Level I robots could perform 12 percent of the 14 million production worker jobs in all

6. Ayres and I (Ayres and Miller, 1983) had previously extrapolated the results of the original 16 survey responses to other metalworking occupations. In Miller, 1983, the extrapolation to other production worker occupations in metalworking was still based on subjective judgment of occupational similarities. The difference is that a more complete set of production worker occupations (given in the 1980 *Occupational Employment* survey [Bureau of Labor Statistics, 1982] was considered in Miller, 1983, than was considered in Ayres and Miller, 1983.

Table 2-13: Revised Estimated Potential Displacement of Production Workers in Industries SIC 34–37

Occupation	Total Production Employment, 1980	Number of Workers Potentially Displaced by:		Potential Percent Displacement	
		Level I	Level II	Level I	Level II
Tool handlers	383,490	113,173	190,974	29.5	49.8
Metalcutting machine operators	1,073,380	153,420	437,660	14.3	40.8
Metalforming machine operators	337,841	110,554	207,193	32.7	61.3
Other machine operators	785,660	126,729	287,446	16.1	36.6
Assemblers	974,410	86,722	280,630	8.9	28.8
Inspectors	216,800	16,260	76,747	7.5	35.4
Material handlers and laborers	479,960	15,887	197,270	3.3	41.1
Miscellaneous craftworkers	507,980	2,624	31,234	0.5	6.1
Maintenance and transport workers	341,740	0	0	0.0	0.0
Total	5,101,261	625,369	1,709,154	12.3	33.5

manufacturing industries (SIC 20–39), and that Level II robots could perform 33 percent of these same jobs. While I acknowledge that production tasks are not really so homogeneous across industries as would be implied by this extrapolation, my rationale for it is that it places an upper bound on the percentage of jobs in all manufacturing that could be robotized. The result of this estimate is that over 1.7 million workers could be displaced by Level I robots and nearly 4.7 million workers by Level II robots. These estimates correspond, respectively, to 1.7 percent and 4.7 percent of the total number of jobs in the civilian work force as of 1980.

For purposes of comparison, recall that the Eikonix (1979) study estimates a potential displacement of 1.2 million workers by Level I robots and 4.4 million workers by Level II robots. The earlier Ayres and Miller (1983) study estimates a potential displacement of 1.5 million and 4.0 million workers by Level I and Level II robots. The 1983 Hunt and Hunt study estimates that, at most, 200,000 workers by 1990 will be displaced by the types of Level I robots in use as of 1983. The magnitude of displacement estimated in the Eikonix, Ayres and Mil-

Table 2-14: Estimated Number of Production Workers Potentially Displaced by Robots in Metalworking Industries (SIC 34–37) and in All Manufacturing Industries

Industries	Total Production Employment, 1980	Number of Workers Potentially Displaced by:	
		Level I	Level II
Metalworking, SIC 34–37	5,101,261	625,369	1,709,154
All manufacturing, SIC 20–39	14,190,289	1,745,405	4,735,747

ler, and Miller (1983) studies, which consider a "long term" horizon (15 or more years), is nearly 10 times larger than the magnitude of displacement estimated in the Hunt and Hunt study, which considers a specific date in the "short term."

The results in Tables 2-13 and 2-14 are also compared with the results of two other studies which have published conclusions, but which are not reviewed here in detail. Volkholz (1982) estimates that in West Germany, industrial robots will displace about 10 percent of the production workers in the metalworking industries sometime between 1990 and 1994, "assuming that the market potential is fully exploited." Volkholz's estimate of the market potential for robot use is based on a consideration of what jobs Level I-type robots are technically capable of performing, as well as a consideration of the costs and benefits of installing robots in those jobs. The estimate of 10 percent displacement of production workers is based on the assumption that the potential market for Level I-type robots in West Germany is 70,000 units, and that by 1990, 60 percent of this potential will be realized (or that about 39,000 units will be installed). Volkholz's estimate of the percentage of production workers that will be displaced by Level I robots in West Germany's metalworking industries by 1994 is about the same as the Miller (1983) study estimate for the United States (Table 2-13).

Porter et al. (1985) estimate that by the year 2000 about 1.3 million workers will be displaced by Level I and Level II robots.[7] Alan Porter (pers. comm., 1986) explained that to derive these estimates, they "subjectively considered how high a percentage of jobs could be eliminated by robots for each fine-grain BLS employment category. The tally resulted in the ballpark estimate of 1% of the civilian work force, or 1.3 million jobs." This is also similar to the Miller study estimate of the number of production workers that could potentially be displaced by Level I robots (Table 2-14). The results of the Porter et al. study suggest that the results for Level I displacement be used as an upper bound for estimated long-term potential displacement resulting directly from both insensate (Level I) and rudimentary sensor-based (Level II) robots. Porter et al. also estimate that by 2000 nearly 10 times more people will be displaced by heavy robotization and other forms of automation than will be directly displaced by robots alone. This would suggest that while the Level II estimate given in Table 2-14 is too high, even assuming the use of sensor-based robots, it is too low for considering the displacement impact of automation in general.

The number of robots that would be in use if the full potential for displacement were to be realized (Table 2-13) is shown in Table 2-15. Even if it were

7. Porter et al. (1985) define four levels of robots: variable sequence (Level I), playback (Level II), numerical control (Level III), and intelligent (Level IV). Levels I, II, and III corespond to the term Level I in the present discussion. Level IV in Porter's classification corresponds to the way the term Level II in this discussion. Their estimated job displacement through the year 2000 assumes the use of all four types of robot: variable sequence, playback, numerical control, and intelligent.

Table 2-15: Level I Robot Population Implied by Estimated Potential Displacement in Metalworking Industries (SIC 34–37) and in All Manufacturing Industries

	Number of Level I Robots, Assuming One Robot Displaces:		
Industries	2 Workers	3 Workers	4 Workers
Metalworking, SIC 34–37 (0.62 million workers displaced)	313,000	208,000	156,000
All manufacturing, SIC 20–39 (1.7 million workers displaced)	873,000	582,000	436,000

assumed that robot use will be confined principally to the metalworking industries, the implication is that from 160,000 to 300,000 Level I robots would be required. To realize the full potential for Level I displacement across all manufacturing industries would require from 440,000 to 870,000 robots. As shown earlier in Table 2-7, recent projections of the U.S. robot population in 1990 estimate 100,000 or fewer units. In the updated 1985 UM/SME Delphi survey, most estimates of the 1995 robot population ranged between 60,000 and 120,-000 units. The conservative and intermediate forecasts given in the Eikonix study, which so far have bounded the actual robot population, predict 56,000-123,000 robots in use in the United States by 1995.

What is evident is that estimates of displacement based on assessments of the proportion of jobs that could be robotized are many times higher than the estimates based on the actual number of robots in use. For example, the estimate of potential displacement by Level I robots across all manufacturing shown in Table 2-14 is 9-15 times greater than Hunt and Hunt's (1983) estimates of actual displacement by robots by 1990. Even if Hunt and Hunt had assumed a larger robot population (say, 50 percent larger), or extended their time horizon several years, it would still be the case that there would be a large gap between the estimates of potential and actual robot use and between the respective displacement estimates. Either the estimates of potential robot use are much too high, or there are factors restraining actual robot use from achieving its full potential. My argument is that estimates of potential displacement are not suspect, and that the key to reconciling the estimates is to explain why many firms that could potentially use robots will not have a strong enough economic incentive to do so.

2.7. The Reliability of the Estimated Potential Displacement

In the discussion comparing the Hunt and Hunt (1983) displacement estimates with other estimates (also shown in Table 2-10), I noted that the original 1981 CMU estimates of potential displacement are comparable to the 1982 Delphi survey estimates of potential displacement for that same year. These estimates of potential displacement were based on data collected from manufacturers over five years ago.

Given increased industrial experience with robot use and further develop-

Table 2-16: Comparison of Revised CMU Survey and UM/SME Delphi Survey Estimated Percentage of Displacement by Occupation

Occupation	Revised CMU Survey Data for Potential Level I Displacement	UM/SME Delphi Survey Results for Actual Displacement by 1995	
		Updated Survey, 1985	Original Survey, 1982
Assemblers	8.9	10	12
Inspectors	7.5	6	15
Packers	10.8	5	10
Painters	43.5	20	20
Welders and flamecutters	25.5	20	15
Machinists and machine operators	14.3	13	10
Tool and die makers	1.3	2	—

ments in technology, one questions whether such numbers are still meaningful indicators. Presenting my revised survey results (Miller, 1983) at this point would be of little value if a more recently completed survey were to have produced substantially different results. To address this concern, the revised estimates of potential displacement from my study are compared in Table 2-16 with the results of the updated UM/SME Delphi survey published in 1985 (Smith and Heytler, 1985). The first column in the table shows my revised estimate of potential displacement by Level I robots, based on data from the 1981 CMU robot survey (shown in Table 2-13). The second column shows the median response as reported in the 1985 Delphi survey to the directive, "Given the current labor content of production, forecast the *actual* percentage of workers that will be displaced in the United States, by occupation, by the year 1995."[8] The third column shows the median response of the forecast of *actual* displacement by occupation, by 1995, as reported in the 1982 Delphi survey.[9]

For the two largest occupational groups—assemblers, and machinists and machine operators—the revised CMU results and the 1985 Delphi results are in close agreement. The results are also very close for inspectors, welders, and tool and die makers. The CMU results are nearly double the 1985 Delphi results for two occupations—packers and painters. The overall conclusion from the comparison is that the more recent Delphi survey produces estimates of displacement that are very similar to the results obtained by Ayres and myself over five years ago.

The differences between the results of the 1982 and 1985 Delphi surveys are interesting to note. For two occupations—welders and machine operators—the

8. Between 30 and 60 people composed the panel that responded to this directive (Smith and Heytler, 1985: 3)

9. The 1985 Delphi survey did not ask for an estimate of potential displacement separate from actual displacement, as it had in 1982. Therefore, the Delphi results for actual displacement must be used for comparison purposes.

median estimates of displacement have increased. For three occupations—
assemblers, inspectors, and packers—the median estimates have decreased.
The median estimate for painters has remained the same. As a result of additional experience with robot use, robot users, producers, and experts have
revised their opinions regarding the actual level of displacement that will be
realized within specific occupations by 1995, but it is clear there has not been a
major change in opinion.

An important change in opinion between the initial and updated Delphi surveys which is not reflected in this comparison of displacement estimates relates
to the perceived impacts of robots in particular versus automation in general.
Some of the comments by respondents to the directive regarding displacement
in the 1985 Delphi survey (Smith and Heytler, 1985: 169) are as follows:

— Answers are with respect to *robots* only; flexible manufacturing systems
  (FMSs) will displace many more.

— Many of the above positions will be eliminated by other forms of automation and technology (optical recognition, real time gauging, etc.).

— The easier applications have been done and in other areas hard automation
  will be more appropriate than robots.

— These answers are for robots only, which will be a relatively minor factor
  affecting labor displacement.

In the 1982 Delphi survey (Smith and Wilson, 1982: A-50), a similar directive elicited no comments of any type.

What do these comments mean for the interpretation of the updated displacement estimates in the Delphi survey? They could mean that the estimates for
displacement due to robot use represent only a fraction of the actual displacement effect that will result from the full range of factory automation by 1995. Or
they could mean that the displacement impacts include the impacts of using
robots as well as other forms of automation. The Delphi survey specifically
stated that respondents were to consider the displacement impacts of robots
only, not the impacts of other forms of automation. Nonetheless, it is not necessarily clear what people intended when they responded to the survey. Also, there
are reasons why estimates of the number of workers displaced by robots will not
necessarily coincide with estimates of the number of robots in use. Namely,
people may not be able to isolate the displacement effects of robots in particular
from automation in general, especially when robots are used as parts of large
automation systems. Issues related to the work-force impact of robots in particular versus factory automation in general are discussed in the concluding section
of this chapter.

2.7.1 ESTIMATED POTENTIAL FOR ROBOTIZED OPERATION OF METALCUTTING MACHINE TOOLS

A different approach is discussed here to check the reliability of the revised estimates of potential displacement. For one application area, the operation of metalcutting machine tools, I have made an estimate independent of the CMU survey-based estimates for the potential for robot use. (The procedure for estimating the percentage of metalcutting machines in the metalworking industries that could be operated by Level I and Level II robots is explained in detail in Appendix D.) In brief, I have assigned 95 of the metalcutting machines included in the "12th *American Machinists* Inventory of Metalworking Equipment" (*American Machinist,* 1978) to four categories:

1. machines which are designed to process low volumes of small to moderate-sized parts;
2. machines which are designed to process any size part under fully automatic control without operator intervention;
3. machines which are designed to process low volumes of very large and/or very heavy parts;
4. machines which are designed to process medium and large volumes of small to moderate-sized parts.

The results of this analysis are compared in Table 2-17 with the revised estimates of the percentage of metalcutting machine operators in the metalworking industries that could be displaced by robots. The estimates of the percentage of metalcutting machine tools that could be operated by Level I robots encompasses the CMU survey-based estimate of the percentage of metalcutting machine tool operators that could be displaced by Level I robots. The estimated percentage of metalcutting machine tools that could be operated by Level II robots is moderately higher than the survey-based estimated percentage of machine tool operators that could be displaced by Level II robots. While these

Table 2-17: Comparison of Percentage Estimates of Potential Robot Use in the Operation of Metalcutting Machine Tools

Source	Level I	Level II
Derived from analysis of metalcutting machine tools in the "12th *American Machinist* Inventory of Metalworking Equipment"[a]	9.4–15.7	46.7
Derived from 1981 CMU survey of current and prospective robot users[b]	14.3	40.8

[a]Figures are estimated percentage of all metalcutting machine tools in all of the metalworking industries that could be operated by a Level I and a Level II robot.

[b]Figures are estimated percentages of metalcutting machine tool operation jobs which could be performed by a Level I or a Level II robot.

two estimates of the potential for robot use were derived independently of one another, they are in close agreement. It appears that there is no reason to disbelieve the survey-based estimated potential for using robots to operate metalcutting machine tools.[10] It is conceivable that respondents to the survey derived their estimates from heuristics that were similar in concept to the ones used in Appendix D to estimate the proportion of metalworking machine operations and other tasks that could be performed by robots.

The estimates of potential displacement for Level I robots derived from the 1981 CMU survey are similar to the updated (1985) UM/SME Delphi survey estimates for actual displacement for the year 1995. Also, for one important applications area, metalcutting machine tool operations, an independently derived estimated potential for robot use is similar to the result derived from the survey data. For these reasons, I argue that the estimates of the potential for Level I robot use are reliable.

The preceding analysis of the proportion of metalcutting machine tools that could be operated (e.g., loaded and unloaded) by a robot implies that, as of about 1980, Level I robots were capable of operating nearly 270,000 machine tools. Yet, Hunt and Hunt (1983) forecast that by 1990, there will be, at most, 42,000 robots used to load and unload all types of machines (metalcutting, metalforming, as well as other types). As noted earlier, it is evident that the Hunt and Hunt forecast, as well as other forecasts of the market for robot use, assumes that only a small proportion of the potential number of robot applications will actually be realized. Identifying reasons why only a fraction of the potential for robot use will be realized over the next decade or so is a starting point for understanding the inconsistency between the estimated potential displacement and the estimated number of robots that will actually be in use.

2.8. Explaining the Gap between Potential and Actual Levels of Robot Use

Hunt and Hunt (1983: 54) "expect diffusion of robotics technology to be limited primarily to large firms, and perhaps even Fortune 500 firms, for the

---

10. Similar analyses of the potential for robot use in other application areas have not been carried out. It is noted that Funk (1984) analyzes the cost of manual and robotic assembly across metalworking industries (SIC 34–38). He reports that Level I robots could be economically justified to replace between 20 and 30 percent of the assembly workers in SIC 34–38. The range represents different assumptions regarding the amount and type of part-positioning equipment required for the Level I robot. This is nearly two to three times the estimate of potential displacement derived from the 1981 CMU survey data. Funk also estimates that Level II robots could be economically justified to replace between 44 and 53 percent of assembly workers in these same industries. Again, the range represents various assumptions regarding the sophistication of the sensor-based systems used. Funk's definitions of Level II robot technology are much more specific than those used in the 1981 CMU survey. His results for Level II robot displacement are about two times higher than the CMU survey-based estimate.

foreseeable future." Certainly early use of robots was concentrated within large firms. This can be seen by examining the background of the respondents to the 1981 CMU robotics survey. Respondents classified themselves according to size of establishment (number of production workers) and mode of production (custom and small batch, medium batch, and mass). The number of respondents within each category is shown in Table 2-18. While current and prospective users of robots are almost evenly spread across small-batch, medium-batch, and mass production, 33 of the 52 respondents are in companies with 1000 or more production workers.[11] As of 1981, current and prospective users of robots were principally concentrated in very large establishments, and many of these large establishments were also part of the largest manufacturing firms in the United States.[12] More recent studies have reported that robot use is highly concentrated in a relatively small number of establishments.[13] Hunt and Hunt apparently believe that for the next 5–10 years, robot use will continue to follow similar patterns.

If robot use over the next decade were to be concentrated within large establishments (which presumably belong to large firms), the inconsistencies between estimates of potential displacement and market forecasts of the robot population would be reconciled. According to data in the *1977 Census of Man-*

11. A total of 52 firms responded to the 1981 CMU robotics survey. However, only 24 of the respondents completed the section on the percentage of workers within selected occupational categories that could be replaced by Level I and Level II robots.

12. Data from the "13th *American Machinist* Inventory of Metalworking Equipment" (American Machinist, 1983) does not support the assertion that, as of the early 1980s, robot use was highly concentrated in the largest establishments. According to the 13th *American Machinist* Inventory, as of early 1983, 57 percent of the robots in use in the metalworking industries were located in establishments with fewer than 500 employees. However, the data for the number of programmable robots in use are suspect. The inventory reports a 1983 U.S. robot population of 2744 units. Other reputable estimates of the 1983 robot population are several times that figure. According to the Robotic Industries Association (1985), there were 6,300 robots in use as of January 1983 and 9400 robots in use at the end of that year. Also, the 13th *American Machinist* Inventory reports that there were only 124 robots in use throughout the entire U.S. motor vehicle industry (SIC 371) in 1983. This figure seems unreasonably low given that automotive manufacturers responding to the 1981 CMU robot survey reported a total of 941 robots in use as of the spring of 1981.

13. Meyer (1985: 813) states: "Although robots are used in almost every industry and type of application, the majority of installations are concentrated in a relatively few plants and types of applications. For example, it is estimated that just 10 plants contain nearly one-third of all robot installations and that the three categories of welding, material handling and machine loading account for approximately 80% of all current applications. At the same time, it must be remembered that the market penetration of robots has been relatively limited in even the most common applications."

The results of a recent survey of people familiar with robot use from business, government, and academics (Industrial Relations Counselors, 1985) provide supporting evidence that, to date, robot use in the United States has been highly concentrated within the largest companies. Over 83 percent of survey respondents indicated that by 1990 companies with over 10,000 employees will be (or already are) significant users of robotics, whereas only 7 percent of respondents believed that companies with fewer than 1,000 employees will be significant users.

Table 2-18: Number of Respondents in the 1981 CMU Robotics Survey by Batch Size and Size of Establishment

Batch Size	Size of Establishment (number of production workers)				Total Number of Respondents
	1–99	100–499	500–999	≥ 1000	
Custom-small batch	2	2	1	11	16
Medium Batch	1	4	5	8	18
Mass	—	3	1	14	18
Total number of respondents	3	9	7	33	52

ufactures (Bureau of the Census, 1981a), almost 40 percent of production workers in SIC 34–37 are employed in large establishments (those with 1000 or more production workers). Across all manufacturing industries (SIC 20–39), only 25 percent of production workers are employed in large establishments. Supposing the estimated potential for Level I robots were to be fully realized, but only within large establishments, then of the 40 percent of manufacturing workers employed in establishments with 1000 or more workers in industries SIC 34–37, 12 percent would be displaced by Level I robots. Assuming one robot displaces two workers and using employment data for 1980 implies a robot population of 122,000 units. A similar calculation, assuming that robots are used across all manufacturing industries, implies a robot population of about 213,000 units. These totals are close to several of the market forecasts of the cumulative robot population for the early to mid-1990s.

Why hypothesize that robot use in the United States will be concentrated in the largest manufacturing establishments belonging to the largest manufacturing firms? Consider the factors given by Hunt and Hunt (1983) to justify their prediction that robot use will be less rapid than indicated in other forecasts they examined. They refer to (1) the lack of trained personnel, (2) the large financial commitment needed, (3) the strong management commitment needed, and (4) general economic conditions which will restrain investment. Arguably, large firms could more easily overcome the first and second factors than small firms because of greater reserves of human and financial resources. However, there is no obvious argument giving large firms an advantage over small firms with respect to the third and fourth factors. While these four factors adequately explain restraints on the growth of the total robot population, they do not fully support the premise that robot use will be primarily limited to large firms.

The hypothesis I put forward here is that large firms have a substantial cost advantage over small firms in using robots. The underlying reason is that the total cost per robot application decreases as the number of applications increases, and therefore firms that can install multiple robots can do so less expensively (on a per-unit basis) than firms that install only one or several units.

Since large firms are more likely to have more opportunities for using robots than small firms, they can more likely buy multiple units and consequently realize a lower per-unit investment cost. In fact, it is argued that for certain types of applications, relatively short payback periods can be realized only if multiple units are installed. In summary, it is my argument that firms which can install multiple robots will realize shorter payback periods than firms that cannot, and will therefore have a stronger financial incentive to install robots where it is feasible to do so. This is one explanation for the concentration of robot use within large firms.

2.9. An Analysis of Economic Incentives for Installing Robots

An illustrative analysis of the costs and benefits of substituting robots for production workers is presented here. The objective of this analysis is to identify under what conditions there would be a strong economic incentive to install robots, given that there is a technical potential for doing so. The basic approach is to calculate the simple payback period, based on realistic estimates of the cost of installing robots and on total labor cost saved. Special attention is given to identifying those conditions under which payback periods based on direct labor savings would be relatively short (three years and less). It is assumed that if the payback periods are relatively short, there will be a strong economic incentive to install the robot, and if they are long, there will not necessarily be a strong enough economic incentive to do so.

2.9.1. ROBOTS USED TO OPERATE METALCUTTING MACHINES
The cost of installing robots to operate metalcutting machine tools is now considered in detail. I have estimated the total cost of implementing a robot to operate metalcutting machine tools considering both robot hardware cost and development cost. Based on the actual cost of various robot models, typical figures for a low-cost, medium-cost, and high-cost robot are specified and are used for the subsequent analysis. Specific models of Level I robots which would typically be used for the loading and unloading of metalcutting machine tools are shown in Table 2-19, and some of the important technical features of these robots (reach, load, and type of control) are shown in Table 2-20.[14]

Low-cost robots have limited reach and payload and are usually used to load/ unload small parts on a single machine tool. It is assumed that a low-cost robot will replace at most one operator per shift because of the limited reach. Because of the very small payload, there would be relatively few applications for low-cost robots in machine operation applications. A medium-cost robot would be

14. While these tables were initially compiled in 1983, price ranges in more recent catalogues of robot information show that the 1983 prices are still reasonable for use in the analysis presented here.

Table 2-19: Typical Robot Models Used to Operate Metalcutting Machine Tools

	Robot Cost[a] (in 1982 dollar values)		
Robot Model	Minimum	Typical	Maximum
LOW-COST RANGE			
Copperweld, CR5	15,000	17,000	20,000
ASEA, MHU junior	—	20,000	—
General			
Numeric, MHO	—	25,000	—
MEDIUM-COST RANGE			
Prab 4200	30,000	35,000	50,000
Unimate 2000	46,000	—	64,000
Prab, E	51,000	62,000	72,000
Unimate 4000	64,000	—	74,500
Prab FA	65,000	80,000	90,000
HIGH-COST RANGE			
ASEA, IRb-60	60,000	75,000	120,000
CM T3 586	75,000	85,000	—
Bendix, ML series CNC	—	100,000	—
Prab, FC	120,000	130,000	150,000

[a]Base price of robots from Flora, 1982.

required to serve a single machine if the typical payload were too heavy for the smaller, low-cost robots. Because they have a longer reach, the medium-cost robots can sometimes be used to load/unload two machine tools. It is assumed that a medium-cost robot will replace from one to two operators per shift. High-cost robots generally have a longer reach, a larger payload and more sophisticated control capabilities. In most situations, a high-cost robot would be under-

Table 2-20: Key Attributes of Selected Robots

Robot Model	Maximum Load (pounds)	Maximum Horizontal Reach (inches)	Maximum Vertical Reach (inches)	Control type
LOW-COST RANGE				
Copperweld, CR5	5	fixed	2	microprocessor; nonservo
ASEA, MHU junior	35	48	64	microprocessor
General				
Numeric, MHO	22	6	6	microprocessor
MEDIUM-COST RANGE				
Prab 4200	125	82	65	microprocessor
Unimate 2000	300	80	83	hardwired PTP or continuous
Prab, E	100	56	58	drum-mechanical
Unimate 4000	450	115	120	microprocessor
Prab FA	250	78	96	drum or microprocessor
HIGH-COST RANGE				
ASEA, IRb-60	225	102	154	microprocessor
CM T3 586	132	89	84	microprocessor
Bendix, ML series CNC	150	98	165	CNC
Prab, FC	2000	90	114	drum or microprocessor

Sources: Flora, 1982; Productivity International, Inc., 1982.

utilized if it were used to serve a single machine tool, since a low- or a medium-cost robot could perform the same task. When the high-cost robots are used for machine tool loading and unloading, they are typically used to serve two or three machines. Since a high-cost robot might also be used to serve only one machine tool, it is assumed that high-cost robots would replace from one to three machine operators per shift.

2.9.2. TOTAL IMPLEMENTATION COSTS FOR A ROBOT APPLICATION

Total implementation costs are almost always substantially larger than the base purchase price of the robot. Susnjara (1982) breaks down the total investment required to install a robot into the following major categories: (1) robot cost, (2) tooling cost, (3) installation cost, and (4) engineering cost. Hasegawa (1985) breaks down total investment into (1) robot cost, (2) accessories cost, (3) engineering cost, (4) programming cost, (5) installation cost, (6) tooling cost, (7) training cost and related expenses. Thus, in addition to the base price of the robot, there are substantial costs incurred for:

— engineering analysis of job requirements and the evaluation of alternative robot models;
— design and construction of end-of-arm tooling for the robot (grippers, special tools, etc.);
— design and construction of part-orientation hardware (position tables, fixtures, etc.);
— engineering analysis and effort to integrate robot, tooling, and accessory hardware;
— miscellaneous items such as training courses and manuals.

In most installations, the combination of these other application costs is several times the base purchase price of the robot. Tanner and Adolfson (1982) cite an example where the cost of the robot is 44 percent of the total implementation cost for a machine-tending application in the auto industry. Based on interviews with users and vendors, Hunt and Hunt (1983) report that the robot itself usually represents less than 40 percent of the total installation cost. Tech Tran Corporation (1983) reports that for a typical machine-loading application, the robot represents 55 percent of the total cost. However, they consider only robot cost, accessories cost, and installation cost, and do not consider other major costs for factors such as planning and analysis and system integration. Respondents to the 1985 UM/SME Delphi survey report that, as of 1985, robots that are part of machine-tending systems constitute only 40–50 percent of the total system cost. They also report that costs associated with planning, installation, and integration amount to 25–50 percent of the value of the robot system itself (Smith and Heytler, 1985: 93–94). The estimates on robot-system cost proportions given above imply that the base price of the robot alone typically accounts for 27–44 percent of the total implementation cost of a system for machine tool operation.[15]

Other reports from and discussions with experienced robot users indicate that the robot base price can constitute an even smaller proportion of total system cost. Toepperwein and Blackman (1980) show figures for a "typical" robot cost justification where the robot is 23 percent of the total installation cost. Hasegawa (1985) describes an application where a robot used to place parts on an assembly accounts for 16 percent of the total system cost. Ron Potter (pers. comm., general manager of Advanced Manufacturing Systems in Norcrosse, Georgia, 1986), breaks down the total cost of a robot installation into three major components:

1. engineering labor costs—include engineering labor time for applications selection and design, software development, and system implementation;
2. robot—includes the "out-of-the-box" cost of the manipulator and its "as is" control unit, exclusive of end-of-arm tooling;
3. other system-related costs—include labor time and resources for such items as end-of-arm tooling and other accessory hardware (e.g., position tables or fixtures), and for relocating machines, if necessary.

Based on experiences of retrofitting several hundred robot systems into existing manufacturing operations, Potter estimates that the major cost components described above typically constitute the following proportion of total robot implementation cost:

engineering labor	20–40%
robot	20–30%
other system related costs	30–60%

Potter claims that "other system related costs" are a large fraction of the total because of the problems associated with "shoehorning the robot into the job" in situations where the robot is retrofitted into existing applications.

There is generally a trade-off between the amount spent for the robot and the additional amount required to fully implement the system. Since high-cost robots generally have many degrees of freedom, a long reach, a large working envelope, and sophisticated control capabilities, a minimum of accessory hard-

15. The low value, 27 percent, assumes that the robot is 40 percent of the system hardware/software costs, and that planning/integration/installation costs equal 50 percent of the total system hardware/software costs. The high value, 44 percent, assumes that the robot is 55 percent of the system hardware/software costs, and that planning/integration/installation costs equal 25 percent of the total system hardware/software costs.

ware and engineering is required to enable the robot to acquire, load, and unload parts. Since low-cost robots are more limited in their degrees of freedom, reach, working envelope, and control capabilities, it is more difficult to get the part to them, requiring more accessory hardware and engineering to compensate for these limitations. Engelberger (1980: 102) also discusses this trade-off:

> Generally speaking the higher priced robots are capable of more demanding jobs and their control sophistication assures that they can be adapted to new jobs when original assignments are completed. So too, the more expensive and more sophisticated robot will ordinarily require less special tooling and lower installation cost. Some of the pick and place robots are no more than adjustable components of automation systems. One popular model, for example, rarely contributes over 20% of the total system cost.

The assumptions used in calculating total implementation costs for low-, medium-, and high-cost robots are shown in Table 2-21. The base price of the robot is assumed to be 20, 25, and 33 percent of total implementation costs, respectively, for the low-, medium-, and high-cost robots. A summary of assumptions used for total implementation costs for retrofitting low-, medium-, and high-cost Level I robot systems into a factory to load and unload machine tools is given in Table 2-22.

2.9.3. CALCULATION OF PAYBACK PERIODS BASED ON DIRECT LABOR SAVINGS

Adopted here is a cost-benefit framework where the benefits are *narrowly* defined as labor savings, and where costs are the total cost of implementing the robot system. While other benefits (such as more consistent and higher quality processing, increased throughput, and improved conditions for workers moved out of unpleasant jobs) are also realized when robots are used, labor savings are widely regarded as the primary (and often the only) variable to consider. Engelberger (1980: 103) makes this point quite clearly:

> The prime issue in justifying a robot is labor displacement. Industrials are mildly interested in shielding workers from hazardous working conditions, but the key motivation is the saving of labor cost by supplanting a human worker by a robot. So very much the better if a single robot can operate for more than one shift and thereby multiply the labor saving potential.

Table 2-21: Assumptions for Calculating Total Implementation Costs

Type of Robot	Robot Base Price	All Other Implementation Costs as a Multiple of Robot Base Price
Low-cost	$20,000	4
Medium-cost	$60,000	3
High-cost	$100,000	2

Table 2-22: Summary of Cost Assumptions for Retrofitting Level I Robot Systems

Robot Hardware Cost (R)	Development Cost (D)	Total Implementation Cost (I = R + D)	Operators Replaced per Shift
20,000	80,000	100,000	1
60,000	180,000	240,000	1–2
100,000	200,000	300,000	1–3

The respondents to the 1981 CMU survey overwhelmingly ranked efforts to reduce labor costs as their main motivation for installing robots, and other surveys have reported similar results (Whitney et al., 1981; *Industrial Robot*, 1981; Ciborra, Migliarese, and Romano, 1980). Dreyfoos and Stregevsky (1985: 834) note that in aerospace manufacturing the chief payoff for using robots "is a cost saving, usually attained by robotizing a procedure that used large sums of labor, for example, drilling, assembly, layup, deburring, sanding, inspection, and wire-harness routing."

In the past several years, there has been greater consideration of the impact of new technological systems on quality, inventory, productivity, flexibility, and the ability to innovate, as well as on direct labor costs (Tepsic, 1983; Kaplan, 1983, 1986). While there is a growing consensus that direct labor savings should not be the primary focus of benefits in a cost-benefit analysis, information from the 1985 UM/SME Delphi survey suggests that, as of 1985, labor savings are still the primary way of quantifying benefits from robot use in actual practice. For nine types of application, respondents were asked to estimate the percentage breakdown of total savings due to a variety of factors including "direct labor productivity" and "product quality."[16] The term *direct labor productivity* refers to labor-cost savings resulting from the direct displacement of labor by robots (Smith and Heytler, 1985: 23). The Delphi survey results are shown in Table 2-23.

In Table 2-23 respondents estimated that 93 percent of the savings currently realized as a result of robot use for machine tool operations (machine tending) are due to an improvement in "direct labor productivity" (i.e., a decrease in direct labor costs). In fact, for all nine application areas shown, respondents reported that the largest share of cost savings is due to improvement in direct labor productivity. The responses for projections of the relative importance of cost savings in 1995 are also shown in the table. While respondents indicate that more of the savings in future years will come from improvement in product quality and other factors, labor savings are still viewed as being important. In fact, respondents project that in 1995, improvement in "direct labor productivity" will still be the single largest reason for cost savings in all nine application

16. Other factors listed in the survey are indirect labor productivity; energy savings; reduction of material waste, floor space, and inventory; simplified management; worker safety; and "other" factors.

Table 2-23: Breakdown of Estimated Percentage of Total Savings from Robot Use Represented by Direct Labor Cost, Product Quality, and Other Factors for 1985 and 1995

	Estimates for 1985			Estimates for 1995		
Application	Direct Labor Productivity	Product Quality	Other Factors[a]	Direct Labor Productivity	Product Quality	Other Factors[a]
Machine tending	93	6	1	74	15	11
Material transfer	56	6	38	55	12	33
Spot welding	50	32	18	60	18	22
Arc welding	62	15	23	54	27	19
Spray painting	40	34	26	34	28	38
Processing	50	18	32	50	21	29
Electronics assembly	38	38	24	41	32	27
Other assembly	41	26	33	47	13	40
Inspection	55	36	9	46	38	16

Source: Smith and Heytler, 1985: 197–205.
[a]Includes savings as a result of improvement in indirect labor productivity, reductions in energy and floor space requirements, material waste, inventory, and worker safety-related costs, a simplification of management, and improvement of other factors not specified.

areas. The results reported in Table 2-23 indicate that, even though cost savings due to quality improvement and other factors are increasingly recognized as being important, it is still instructive to focus on direct labor savings (especially in machine tool operations) when trying to analyze the financial incentives for using robots.

Based on the number of workers that could be displaced by one robot and on the cost of implementing the robot, a payback period is calculated. The simple payback period, P, is defined here as

$$P = \frac{I}{wL}$$

where   (2.1)

I = total robot implementation cost

w = total cost per worker per year

L = total workers replaced = (workers replaced per shift) × (number of shifts)

Payback periods are calculated for the low-, medium-, and high-cost robots and for one-, two-, and three-shift operations in Table 2-24. Payback periods are calculated on the assumption that the total cost (wages plus benefits) for a machine operator is $25,000 per year.[17] The table is read as follows: If the

17. Ostwald (1981: 12–27) has compiled direct hourly wage rates (without benefits) for many types of machine operators. His figures are compiled from the Bureau of Labor Statistics and other sources. A sample of the hourly wage rates in Chicago for mid-1982 are as follows: turret lathe operator, $9.57; automatic screw machine, $10.41; bed milling, $9.52; upright drilling, $8.59;

Table 2-24: Payback Period (in Years) for Level I Robots, Based on Labor Savings

Replacement Rate per Shift	Number of Shifts per Day		
	1	2	3
1 robot : 1 worker			
$20,000 robot	4.0	2.0	1.3
$60,000 robot	9.6	4.8	3.2
$100,000 robot	12.0	6.0	4.0
1 robot : 2 workers			
$60,000 robot	4.8	2.4	1.6
$100,000 robot	6.0	3.0	1.9
1 robot : 3 workers			
$100,000 robot	4.0	2.0	1.3

Labor savings per worker replaced are assumed to equal $25,000 per year. The robot costs shown are base prices of robot hardware. For the corresponding total costs of robot application, see Table 2-22.

$20,000-base-price robot ($100,000 total cost) replaces one worker per shift for one shift, the payback period will be 4.0 years. If this robot replaces one worker per shift for two shifts, the payback period will drop to 2.0 years. For a given replacement rate the payback period is longer for the more expensive robot. For a given number of shifts worked per day, the $20,000-base-price robot, which eliminates at most one worker, has the shortest payback period. If the $100,000-base-price robot is used to eliminate three workers per shift for one shift, it will have the same payback as the $20,000-base-price robot. Even though the *total* cost of the high-cost system is three times that of the low-cost system, the payback periods are the same, because the high-cost system eliminates three times as many workers. The annual savings required to achieve a specified payback period for a given total investment cost is shown in Figure 2-1.

Given these estimates of the payback period for various conditions, the objective is to identify the conditions for which there would be a strong economic incentive to replace a worker with a robot, given that there is a technical potential for doing so. Both the 1982 and 1985 UM/SME Delphi surveys report that, throughout industry, payback periods of two to three years are required to justify an investment in a robot application (Smith and Wilson, 1982: A-60; Smith and Heytler, 1985: 102–103). For many firms, the cut-off limit might be closer to two years, but it appears that it will seldom be substantially longer than three years. Engelberger (1980: 104) comments: ". . . Most accountants would find no problem at all in approving proposals yielding a payback period of one or two

---

combination machine operators, $9.19. The Bureau of Labor Statistics' Industry Wage Survey for November 1983 reports hourly wages for machine operators in the machinery manufacturing industry (SIC 35) in various cities across the United States as follows: $6–$9 in Atlanta; $7–$13 in Chicago; $8–$12 in Cleveland; $7–$11 in Dallas; $9–$14 in Detroit; $6–$15 in Los Angeles; and $8–$10 in Worcester. If one assumes an average hourly wage rate of $10 per hour, a 25 percent burden rate for benefits, which would increase the hourly cost to $12.50, and assuming 2,000 hours paid per year, the annual cost figure will compute to $25,000.

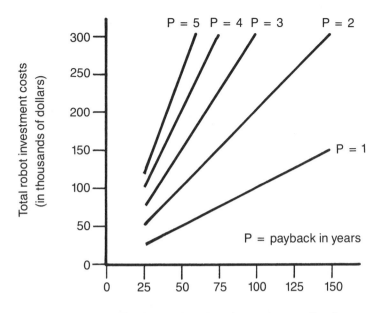

Figure 2-1: Payback Periods for Specified Levels of Total Robot Costs and Annual Savings

years. Beyond that, payback of three or four years could find support." For purposes of discussing the implications of Table 2-24, it is assumed that a robot would be installed only if the projected payback period were to be three years or less. Therefore, no robots (low-, medium-, or high-cost) would be installed in a single-shift operation. If the operation were to run for two shifts per day, attaining the minimal payback period would depend on the type of robot used and on the number of workers it replaced. The minimum payback period would be realized if the low-cost ($20,000 base price) robot could be used. In order to meet the three-year limit, the medium-cost ($60,000) and high-cost ($100,000) robots would have to eliminate two or more workers on both shifts. If the low-cost robot could not be used (because its payload and reach is too small), and if it were possible to eliminate only one job with the medium- or high-cost robot, then the payback period would be nearly five years or more, which is long enough to discourage the application in many instances. Clearly, the shortest payback periods would be achieved if the operation were to run for all three shifts. For the three-shift case, all but one of the longest payback periods shown in Table 2-24 would be short enough to convince management to proceed with the application. (If the robot were to replace only one worker per shift, the payback period for the high-cost robot would be too long, but the period for the medium-cost robot would be just short enough.)

If the three-year payback period were actually a hard and fast rule governing the implementation of robotic technologies (which it is not, of course), one would conclude from this simplified analysis that, where it is technically feasible to do so, the following types of plants would use Level I robots:

— those plants with enough demand to operate on a three-shift basis;

— those plants where it is possible to eliminate two or more workers per shift for two shifts with a medium- or high-cost robot;

— those plants which can use a low-cost robot to eliminate one worker per shift for two shifts.

Taking a conservative outlook, suppose it were the case that one robot eliminates one worker per shift, that paybacks are calculated on a two-shift basis, and that the "heavy duty" robots ($60,000 and $100,000 base price) are required for most machine-loading applications, especially in heavy manufacturing.[18] Given a $50,000 annual savings (two workers at $25,000 per worker), Figure 2-1 shows that total robot costs would have to be $150,000 or less to realize a payback of three years or less. Given the assumptions used here about the total investment required to use a robot, the payback would be nearly five years for the $60,000-base-price robot. The payback period would be even longer for the more expensive robot.

These longer-than-desirable payback periods would discourage some financial analysts from giving the go ahead on robot application. Such payback periods could limit much of the growth in robot use to those firms that have made a high-level policy decision to invest in the technology and to implement robots for reasons that are broader than short-term direct labor savings. For example, Dreyfoos and Stregevsky (1985: 838) state, "For their initial robot applications, many aerospace firms settle for a low return on investment to gain initial production experience with robots." *The conclusion here is that if one takes the most conservative view of the economics of robot use (i.e., robots are viewed only as labor savers and must pay for themselves in a very short time period),*

18. The 1985 Delphi survey results show that as of 1985 a single robot used on a two-shift operation displaces two or three workers (Smith and Heytler, 1985: 184). Respondents expect that by 1995, a single robot used on a two-shift operation will typically displace from three to five workers. In comparison, Hunt and Hunt (1983: 66) state, "Our interviews strongly support the following conclusion about the average displacement effect of robots: one robot replaces one worker per shift." They assume a total of two workers displaced per robot.

then it appears that substantially fewer robots will be installed than could be used, because many potential applications would not meet requirements for short payback periods.

2.9.4. CALCULATION OF PAYBACK PERIODS WHEN MULTIPLE ROBOTS ARE USED

In the preceding discussion, the total systems cost of installing one robot is given by

$$I = R + D \qquad (2.2)$$

where

I = cost to install one robot

R = robot base price

D = development and applications cost

In the payback-period calculations in Table 2-24, it is assumed that only one robot will be installed. It is implicitly assumed that if more than one robot is considered, the payback analysis for installing multiple robots will reduce to the payback analysis of installing one robot. That is, the total cost of installing n robots is n times the cost of installing one robot, or

$$I_n = n(R + D) \qquad (2.3)$$

= total cost of installing n robots

If this were the case, then the number of robots needed to replace L workers is given by

n = number of robots installed = L/d

where

L = number of workers to be replaced

d = workers replaced/robot

Total labor savings is given by (ndw), and the payback is given by

$$P = \frac{(L/d)I}{(L/d)dw} \qquad (2.4)$$

which is equal to the simple formula (Eq. 2.1) used to calculate the payback periods in Table 2-24.

Robot applications engineers and consultants point out that in fact the cost of installing n robots *is not* simply given by n times the cost of a single installation. Actual data on price quotations for robot installations supplied by a robot vendor in the U.S. (shown in Figures 2-2 through 2-4), illustrate this point. The data in the figures are derived by taking the total price quoted by the vendor for an application and dividing it by the number of robots included in the quote. Data points in each figure represent the average price per robot.

The average price per robot for spot-welding applications is shown in Figure 2-2. For an order of a single spot-welding robot, the total price quoted was $90,000. For an order of 103 robots, the total price quoted was about $8 million,

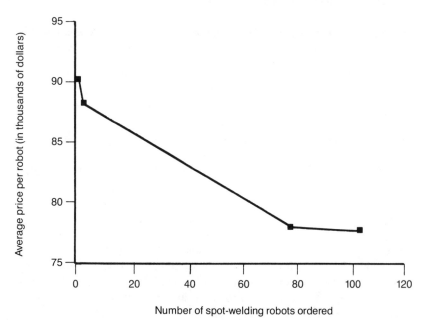

Figure 2-2: Price Quotations for Spot-Welding Robot Applications from Robot Vendor

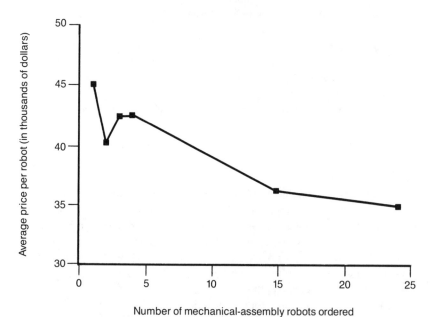

Figure 2-3: Price Quotations for Mechanical-Assembly Robot Applications from Robot Vendor

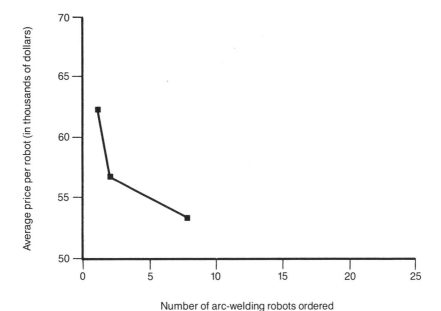

Number of arc-welding robots ordered

Figure 2-4: Price Quotations for Arc-Welding Robot Applications from Robot Vendor

or $78,000 per robot. For the spot-welding applications shown in Figure 2-2, the price quote from the vendor consists almost entirely of the cost for the robot itself. Planning, engineering, system integration, and other systems-related costs are not included in the price quotes, since the automotive companies, the primary users of spot-welding robots, do most of this work themselves. The chief reason for the decrease in the average price quoted per robot is that the vendor is providing a discount on larger orders to "sweeten" the deal and win the contract.

Data on the average price per robot for mechanical-assembly and arc-welding applications are shown in Figures 2-3 and 2-4. In both types of application, the vendor is providing some of the required applications engineering effort and accessory hardware, as well as the robot itself. According to a marketing analyst for the robot vendor, the downward trend in the average price quoted per robot is the result of two factors. The first factor is that the applications engineering and design of accessory tooling can be applied to the second and subsequent units for an order of multiple robots that will be used in similar ways. When application plans can be modified without having to start from scratch, the development costs for the second and subsequent units are lower than for the first unit. The second factor is that the vendor offers a discount on the "out-of-the-box" robot price for larger orders. However, until present time, most of the larger orders (say 50 or more robots) have been for spot-welding or spray-painting applications. Note that the maximum number of units in one order for the mechanical-assembly (24) and arc-welding (8) applications is much smaller than for spot welding (103). The market analyst for the vendor claims that for the cases represented in Figure 2-3 and 2-4 the first factor, applications engineering and design, exerts greater influence than the second factor on the downward trend in the average price per robot when multiple units are ordered.

Figure 2-5, price quotation data for material-handling applications from the same vendor are included to show that the trend in average price per unit versus the number of units ordered does not always exhibit a downward trend. For these applications, it appears that the average price quoted per robot actually *increases* as more units are purchased. What makes this trend different from the others is that these applications include much more "systems-related" work than the applications shown in Figures 2-2, 2-3, and 2-4. The material-handling applications include large expenditures for conveyors and other material-handling hardware in addition to the robotic manipulator and hardware directly related to the robot (e.g., fixturing and end-of-arm tooling). These price quotations more appropriately represent the case where material handling is automated, as opposed to the case where an application is "robotized." The total price of the application divided by the number of robots used does not exhibit the downward trend because such a large fraction of the cost in these applications is unrelated to the implementation of robots.

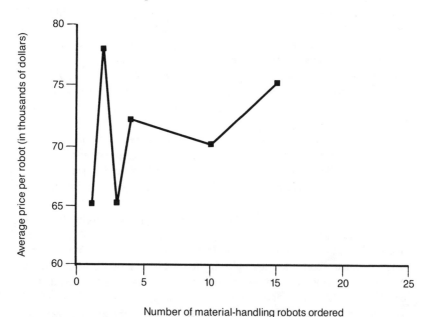

Number of material-handling robots ordered

Figure 2-5: Price Quotations for Material-Handling Robot Applications from Robot Vendor

Based on experience with retrofitting robots into existing manufacturing operations, Potter (pers. comm., 1986) estimates that the robotic manipulator itself typically constitutes 25 percent of the total cost of the application, and that engineering labor and other systems-related costs compose the remaining 75 percent. Potter notes that for some types of application work done by his company, a rough guideline used in estimation is a decrease of about 10 percent in engineering and systems-related costs for each successive application in a similar installation. A similar installation in this context means that applications engineering and the design of all accessory tooling and hardware for the second and subsequent units can be generated by modifying the original plans, without having to start from scratch.[19]

Since only one type of robot application, the operation of metalcutting machine tools, is considered in this analysis, it should be assumed that the development cost for the second robot and for subsequent installations within a particular plant is less than for the first installation. Thus, Equation 2.3 and the payback periods in Table 2-24 would be valid only if n robots were to be installed in n different plants, or if n different types of robot application were installed in one plant. If there were to be more than one installation of a similar application within a particular plant, then the payback periods in Table 2-24 would be upper limits, since they are computed under the assumption that there is no cost savings for multiple installations.

The total investment cost required to install a robot application is now recomputed in a way that acknowledges the decline in the marginal cost per installation for a user who installs multiple units in a similar application. There is no standard formula for this calculation. The market analyst from the robot vendor who provided the data in Figures 2-2 through 2-5 claims that, while there is a consensus that the marginal cost per unit decreases for multiple unit orders, there are no general rules used by costing engineers within the company to estimate these costs. While Potter also claims that it is difficult to generalize about how costs will decrease, his general guideline that a decrease of 10 percent or so will occur for each successive installation is used for this illustrative analysis to capture the phenomenon of decreasing marginal cost per unit. Using this general guideline, if two similar applications of robots operating metalcutting machines were to be installed, total installation cost would be approximated by

$$I_2 = 2R + D + (1 - a)D$$

where the parameter a is the percentage of reduction in the development portion

19. If one robot were being installed across a variety of applications, such as welding, assembly, the operation of metalcutting and metalforming machines, and painting, then it would not necessarily be the case that the development of one type of application will result in a savings in the development of another type. For example, the tooling required for welding is completely different from that required for assembly. Developing the tooling for one application will not reduce development costs for the other.

of the cost (all costs in addition to the base price of the robot) for each successive installation. Generalizing the notion of a constant percentage decrease in the development cost for each successive installation of a similar application, the total cost of installing n robots in an establishment is given by

$$I_n = nR + D \sum_{i=0}^{n-1} (1 - a)^i \tag{2.5}$$

It is acknowledged that this is a highly simplified representation of a more complicated phenomenon. The results of Equation 2.5 are displayed in Figures 2-6 and 2-7. Three values for the percentage of decrease in the development portion of total implementation costs (parameter a) are shown: $a = 5, 10,$ or 20 percent. Figure 2-6 shows cost as a function of the number of robots installed for the $60,000-base-price robot. Figure 2-7 shows similar results for the $100,000-base-price robot. What happens to payback periods if the cost of installing multiple similar robot applications is approximated by Equation 2.5 as opposed to Equation 2.3? They are shorter, clearly, since the investment cost decreases but the labor savings are the same. However, the extent of the decrease depends on how many robots are installed per establishment.

The average number of robots for machine-operating applications that could be installed per establishment in industries SIC 34–37 is estimated as described below. Table 2-13 gives estimates of the number of metalcutting machine operators that could be displaced by Level I robots. With about 1 million workers in this category and a potential displacement estimate of about 14 percent, the

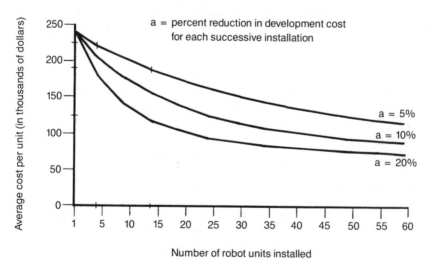

Figure 2-6: Average Cost per Robot for Multiple Installations with Similar Applications: Robot Base Price = $60,000

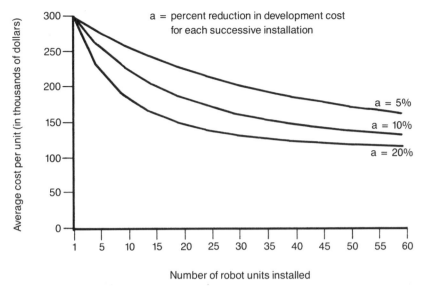

Figure 2-7: Average Cost per Robot for Multiple Installations with Similar Applications: Robot Base Price = $100,000

result is an estimated potential displacement of about 153,000 workers. The 153,000 workers are assumed to be distributed in different size classes of establishments. It is assumed that the number of workers that could be displaced within establishments of a given size class is proportional to the percentage of production workers employed within establishments of that size class. For example, in SIC 34–37, 39.0 percent of all production workers are employed in firms with 1000 or more production workers. (Data on the distribution of employment by size of establishment are taken from the *1977 Census of Manufactures* [Bureau of the Census, 1981a].) It is assumed therefore that 39.0 percent, or 59,834, of the 153,000 metalcutting operators that could be displaced by Level I robots are located in establishments with 1000 or more production workers. Since there are 1012 establishments with 1000 or more production workers in SIC 34–37, there could be, on average, 59 metalcutting machine operators displaced in each of the large establishments (see Table 2-25). The number of robots required to replace 59 workers depends on the number of workers replaced per robot and on the number of shifts worked. For example, if one robot replaces one worker per shift for two shifts, then 30 robots would be required to replace the 59 workers. If two workers were replaced per robot on three shifts, 10 robots would be required (see Table 2-26).

Based on the assumption that the distribution of potential displacement by size class of establishment follows the distribution of employment, it appears that the small establishments (99 or fewer production workers) in this industry

Table 2-25: Number of Workers Potentially Displaced by Level I Robots in the Metalcutting Machine Operation Occupations in Industries SIC 34–37, by Size of Establishment

Size Class	Percent Employment[a]	Total Number of Workers Displaced	Number of Establishments	Number of Workers Displaced per Establishment
1–4	1.2	1,841	39,736	<1
5–9	1.5	2,455	21,591	<1
10–19	3.3	5,063	16,250	<1
20–49	7.5	11,507	16,260	<1
50–99	8.0	12,274	7,803	1–2
100–249	13.8	21,172	6,157	3–4
250–499	12.9	19,791	2,608	7–8
500–999	12.8	19,638	1,342	15
≥ 1000	39.0	59,834	1,012	59

[a]Data compiled from 1977 Census of Manufactures (Bureau of the Census, 1981c, d).

might use, at most, one robot to operate machine tools. The medium-sized establishments (100–499 production workers) might use from one to four robots. The larger establishments, with 500–999 workers, might use from 3 to 8 robots, and those with 1000 or more workers might use from 10 to 30 robots to operate metalcutting machine tools. Total capital costs and average cost per robot are given in Table 2-27 for installing 1, 15, and 30 robots based on the cost assumptions for multiple installations given above. For one $60,000-base-price robot, the average and total installation cost is $240,000. If 15 of these robots were to be used for similar applications within an establishment, total cost would increase to over $2.8 million (where a = 5 percent) or $2.3 million (where a = 10 percent). Average cost per installation would decrease to $189,000 and $155,000 respectively. If 30 robots were to be installed, total installation cost would climb to $4.6 million (where a = 5 percent) and $3.5 million (where a = 10 percent), but average cost per installation would drop to $154,000 and $117,000, respectively. If, in fact, the development cost for each additional installation of a similar application were to decrease, and Equation 2.5 is a reasonable approximation of total installation cost, then there would

Table 2-26: Number of Level I Robots Required per Establishment of Different Size Classes in Industries SIC 34–37 for Metalcutting Machine Tool Operations

Number of Workers Displaced per Establishment	Number of Robots Required			
	1R:1W:2S	1R:1W:3S	1R:2W:2S	1R:2W:3S
1	1	1	1	1
2	1	1	1	1
3	2	1	1	1
4	2	2	1	1
8	4	3	2	2
15	8	5	4	3
59	30	20	15	10

1R:1W:2S reads as follows: One robot replaces one worker per shift for two shifts.

Table 2-27: Total and Average Cost of $60,000-Base-Price Robot, Assuming Development Cost Decreases with Multiple Applications (in Thousands of Dollars)

Number of Robots	a = 5%		a = 10%	
	Total Cost[a]	Average Cost per Robot	Total Cost[a]	Average Cost per Robot
1	240	240	240	240
15	2,832	189	2,329	155
30	4,627	154	3,523	117

[a]Total cost of installing n robots is approximated by

$$I_n = nR + D\sum_{i=0}^{n-1}(1-a)^i = nR + D(\frac{1-(1-a)^n}{a})$$

where

I_n = total cost of installing n robots
R = robot base price
D = development cost for first installation
a = constant percentage of decrease in development portion of
    total cost for each successive installation
Key assumption: The development cost for each successive application of a *similar* installation decreases by a constant percent (given by the parameter a).

clearly be substantial benefits to those users who could install many robots.[20]

The payback periods for the different size classes of establishments are calculated for different assumptions regarding the rate of decrease in the development portion of costs (Tables 2-28, 2-29, and 2-30). The calculations are derived from the estimates of the number of metalcutting machine operators potentially displaced per establishment, by size class of establishment (Table 2-25), and from the assumption that total installation costs are given by Equation 2.5. Labor savings are calculated on the assumption that total cost (wages plus benefits) per worker is $25,000 per year. Today, the typical case is that one robot replaces one worker per shift and that operations run for two shifts (the first column in each table). For this case, the largest establishments (1000 or more workers) would each install 30 robots, the next largest (500–999) would install 8 robots, and those with 250–499 workers would install 4 robots (Tables 2-25 and 2-26). From Tables 2-28, 2-29, and 2-30, we can see which size establishments could use enough robots to drive down total implementation costs to the point where a payback of three years or less could be realized.[21] If the medium-cost robot ($60,000 base price) were used, only the largest establishments would

20. Engelberger (1980: 110) comments, "It has been said that, for practical and economic purposes, robots are a far better proposition when the robot work force numbers upwards of three or four." He mentions that if there is only one robot, a maintenance facility has to be set up from scratch and fully charged against the one unit. With multiple robots, the cost of the maintenance can be spread over several units.

21. The assumption that the development portion of the total installation cost decreases by 20 percent for each additional unit is viewed by robot manufacturers as being overly optimistic. The payback period calculations based on the assumption that the parameter a = 20 percent are shown in Table 2-30 for completeness. However, the results for this case are not discussed further.

Table 2-28: Payback Periods (in Years) Based on Production Labor Savings by Size of Establishment, Assuming a 5 Percent Reduction in the Development Portion of Total Robot Costs per Successive Installation

Size of Establishment	1R:1W:2S	1R:1W:3S	1R:2W:2S	1R:2W:3S
	$60,000 Robot (base price)			
1–49	9.6	9.6	9.6	9.6
50–99	4.8	4.8	4.8	4.8
100–249	4.7	4.7	2.4	2.4
250–499	4.5	3.5	2.4	2.4
500–999	4.5	3.0	2.4	1.8
≥1000	3.1	2.4	1.9	1.4
	$100,000 Robot (base price)			
1–49	12.0	12.0	12.0	12.0
50–99	6.0	6.0	6.0	6.0
100–249	5.9	5.9	3.0	3.0
250–499	5.7	4.4	3.0	3.0
500–999	5.7	3.7	3.0	2.3
≥1000	4.2	3.1	2.5	1.8

Labor savings per worker displaced is assumed to equal $25,000 per year.
1R:1W:2S reads as follows: One robot replaces one worker per shift for two shifts.

Table 2-29: Payback Periods (in Years) Based on Production Labor Savings by Size of Establishment, Assuming a 10 Percent Reduction in the Development Portion of Total Robot Costs per Successive Installation

Size of Establishment	1R:1W:2S	1R:1W:3S	1R:2W:2S	1R:2W:3S
	$60,000 Robot (base price)			
1–49	9.6	9.6	9.6	9.6
50–99	4.8	4.8	4.8	4.8
100–249	4.6	4.6	2.4	2.4
250–499	4.3	3.3	2.3	2.3
500–999	4.0	2.8	2.3	1.8
≥1000	2.4	1.9	1.6	1.2
	$100,000 Robot (base price)			
1–49	12.0	12.0	12.0	12.0
50–99	6.0	6.0	6.0	6.0
100–249	5.8	5.8	3.0	3.0
250–499	5.4	4.2	2.9	2.9
500–999	5.2	3.5	2.9	2.2
≥1000	3.3	2.5	2.1	1.6

Labor savings per worker displaced is assumed to equal $25,000 per year.
1R:1W:2S reads as follows: One robot replaces one worker per shift for two shifts.

Table 2-30: Payback Periods (in Years) Based on Production Labor Savings by Size of Establishment, Assuming a 20 Percent Reduction in the Development Portion of Total Robot Costs per Successive Installation

Size of Establishment	1R:1W:2S	1R:1W:3S	1R:2W:2S	1R:2W:3S
	$60,000 Robot (base price)			
1–49	9.6	9.6	9.6	9.6
50–99	4.8	4.8	4.8	4.8
100–249	4.4	4.4	2.4	2.4
250–499	3.9	3.1	2.2	2.2
500–999	3.3	2.4	2.1	1.7
≥ 1000	1.8	1.4	1.2	1.0
	$100,000 Robot (base price)			
1–49	12.0	12.0	12.0	12.0
50–99	6.0	6.0	6.0	6.0
100–249	5.6	5.6	3.0	3.0
250–499	5.0	3.9	2.8	2.8
500–999	4.4	3.1	2.6	2.1
≥ 1000	2.7	2.0	1.7	1.3

Labor savings per worker displaced is assumed to equal $25,000 per year.
1R:1W:2S reads as follows: One robot replaces one worker per shift for two shifts.

realize the desired payback period. If the high-cost robot ($100,000 base price) were used, along with more realistic estimates of the development-cost-reduction parameters (i.e., a = 5 percent or 10 percent), payback periods for a given size establishment would be even longer, and not even the largest establishments would realize a three-year payback.

If more workers per robot were replaced, then a larger benefit would be realized for a given size investment and payback periods would be shorter. The remaining columns in Tables 2-28 through 2-30 show payback periods assuming three, four, and six workers to be replaced per robot installed. If three workers were replaced per robot, establishments with 500 or more workers could realize a payback period of three years or less with the medium-cost robot. If four workers were replaced, even firms with 250 or more workers could realize a payback period of three years or less with the medium-cost robot.

In this more detailed analysis of payback periods, it is assumed that more workers will be displaced in the establishments with higher numbers of production workers, and that for similar types of application the average cost per robot will decrease as the number of robots installed increases. Given these assumptions, payback periods in the largest establishments (1000 or more production workers) would be substantially shorter than in all other size classes. With the conservative assumption that only two workers would be replaced per robot, only the largest establishments could realize a payback period of three years or less. It seems that of all the establishments only the largest ones could justify the cost of robotization under the most conservative assumptions. In order for the medium-size establishments (those with 99–999 workers) to realize a reasonably short payback period (three years or less), either medium-cost robots would have to be capable of replacing three (and sometimes more) workers, or the cost for multiple robot installations would have to decline even more steeply (for example, as in Table 2-30, where a = 20 percent). This analysis provides a rationale to support claims made by other analysts (e.g., Hunt and Hunt [1983: 53]; and Yoshikawa, Rathmill, and Hatvany [1981: 63]) that over the next decade robot use will be confined primarily to large firms, and perhaps, even to the largest firms.

2.9.5. COSTS OF INSTALLING LEVEL II ROBOTS

In the preceding analysis of costs and benefits, assumptions regarding the extent of job displacement were based on the CMU survey-based results for Level I (insensate) robots. As pointed out earlier, the Level I estimates are comparable to the UM/SME Delphi survey results for estimated levels of displacement for the year 1995. Estimates of the fraction of workers that could be displaced by Level II robots are nearly three times greater than those for Level I. Given the ambiguity of the definition of "Level II" capabilities in the CMU survey and the difficulty of understanding capabilities of sensor-based robotic systems, the quality of the survey data is suspect, and there is no point in considering the implications of the Level II displacement results in further detail. However, it is still possible and useful to consider how the analysis of Level I costs and benefits given above would be altered if Level II robots were considered.[22]

The first issue to consider is the cost required to install a Level II robot for a machine operation application. For example, consider a robot with a vision sys-

---

22. In the CMU survey, Level II robots were ambiguously defined as "robots with rudimentary sensing capabilities." Respondents did not specify what type of sensing they were considering when they responded, and they did not specify *why* they thought "rudimentary sensing capabilities" would make a difference. Machine vision is the most active area of sensor research, and vision systems are increasing in commercial popularity. Without additional information, it is assumed here that the survey respondents were considering machine vision when they gave the responses for Level II (sensor-based) robots. It cannot be verified, however, that survey respondents understood the relatively restricted capabilities of machine-vision systems. For example, it would be possible for machine vision to increase the number of machine operators that could be replaced by a robot if the problem were acquiring parts whose positions are not precisely known or visually inspecting the part. In most cases, it would not be possible for machine vision to increase the number of metalcutting and metalforming operators that could be replaced by a robot if the problem were setting up a machine and ensuring that all tools and mechanisms are securely fastened, are in tolerance, and are properly working. There is no way of knowing the extent to which respondents assumed that a sensor-based system would enable a robot to solve the latter problem versus the first two problems. For the estimates of potential displacement of machine tool operators by Level II robots to be consistent with the known capabilities of machine-vision systems, it must be assumed that the vision system enables the robot to acquire and/or inspect parts as opposed to enabling the robot to perform many of the tasks involved in setting up a machine.

tem used to assist with part location, acquisition, and inspection. A typical cost of a commercial vision module that could perform this type of task would be in the $30,000 range. The cost of implementing the vision module would be approximately the same as the module itself.[23] Less sophisticated types of robot control systems (typically found on the low-cost robots) are not designed to interface with the input from a vision module. In most cases vision systems are used with the medium- or high-cost robots.[24] The addition of a vision system would add about $60,000 to the hardware cost of a medium- or high-cost robot.

Assuming that the vision system were used for part acquisition and/or inspection, would the medium- or high-cost robot with the commercially available vision-system have the capability to perform three times as many jobs as an insensate robot? The answer is almost certainly no if the only means of communication between the vision-system and robot is the standard interfaces that have been provided to date by the vendors of both the robot and vision-system equipment. In their discussion of advanced sensor-based robotic systems, Sanderson and Perry (1983) state that "most systems in use today are configurations of commercial devices which require extensive effort to achieve efficient communication" (p. 864). They note, "In practice, the design of interfacing capabilities and software protocols are major obstacles to efficient networking of commercial devices and discourage the implementation of many interactive systems" (p. 865).

In their discussion of software systems needed to control machines in automated factories, Bourne and Fox (1984) discuss the technical challenges of controlling a system composed of a collection of different types of equipment, with each type controlled by its own "home-brewed" language. They note that for such a system to operate automatically, software is needed that can perform translations back and forth between the control languages of the different types of equipment, and that messages from different machines must be *understood* so that the overall system controller can respond to error and status reports generated by the separate machines. The authors comment that the development of software that permits a group of machines from different vendors to be controlled as a system "has, for the most part, been side stepped by commercial suppliers of flexible manufacturing systems, because, for marketing reasons, it is in their best interest to sell whole systems. This allows them to standardize around one machine controller that is capable of controlling machines built by

23. According to one research scientist at the CMU Robotics Institute, the vision programming isn't usually very complicated, because the vision module is engineered to make it easy to set up. The environmental engineering—adjusting the lighting and protecting the camera lens—is the only difficult part and has to be done for each application.

24. There are exceptions to this statement. For example, in 1983 Copperweld sold the CR5 robot, a low-cost robot with a very simple vision system used in a very restricted way to detect the absence or presence of something in its path. This is the lowest level of machine vision possible, and can be accomplished with other types of transducers.

the same supplier" (p. 81). They note that even when one vendor builds an entire system, it is designed with a "weak link" between the central computer and individual machine controllers so that all that is expected from the communications system is a "go to the next program" command. Bourne and Fox point out that these limitations "seriously limit the flexibility of so called 'flexible' manufacturing systems" (p. 81).

Information from the 1985 UM/SME Delphi survey indicates that a wide range of robot users are still struggling with integrating sensor-based systems, communications, and controls into advanced robotic systems. When asked to list the technological barriers that represent constraints on the utilization of robots in U.S. industries, the most frequently mentioned factors include lack of nonvision sensory systems, limitations in programming, in the control system, and in communications, and inadequate vision systems. Respondents also believed that limitations with vision systems, robot software, and communication across different types of equipment are among the most severe technological barriers restricting robot use (Smith and Heytler, 1985: 105–107).

The conclusion is that, as of present time, additional expenditures for the purchase and installation of sensing equipment will still not result in a robot system with the capability to perform several times as many operator jobs as the insensate system. The reason is that the sensor-based system requires a sophisticated communication and control system beyond what has been provided to date by the major vendors of automation systems. How much will it cost a robot user to acquire the communication and control systems required to take full advantage of sensor-based systems? Clearly, that depends on whether or not such systems are supplied by the major vendors of automation equipment. Recent experience indicates that if vendors do not sell such systems, and individual users are forced to develop their own "customized" solutions in house, the cost for sophisticated communications and controls could exceed several hundred thousand dollars.[25] If this level of extra development effort were required, a Level II installation would be nearly twice as costly as the Level I installation discussed earlier. Since the Level II robot would replace the same number of workers as the Level I robot, the payback periods for the more sophisticated system would take much longer.

Given the longer paybacks and difficulty of upgrading the communication and control between the vision system and the robot, it is not surprising that very few Level II systems have been installed. Clearly, if there are only limited situa-

25. For example, in the experimental manufacturing cell described in Wright et al., 1982, two full man-years of engineering effort were required to adapt the communication and control processes of the robot/vision system and to develop the necessary software to give the integrated robot/vision system substantially improved capabilities over an insensate robot system. Assuming that a full man-year of engineering and programming time (with overhead support) costs roughly $100,000 in an industrial or a research and development environment, this adds another $200,000 to the cost of the robot/vision system hardware.

tions under which there are strong economic incentives to install Level I robots, there are far fewer situations under which even more expensive applications would be installed.

What would reduce the cost of installing a Level II robot? Eliminating the need for the user to upgrade the communication and control systems would in turn eliminate extra development costs. Special "turn key" integrated vision/robot systems have already been marketed, which demonstrates that for a limited class of applications, commercial systems that eliminate this need to upgrade can, in fact, be built. Standardized procedures for transmitting messages across different types of machines, such as the manufacturing application protocol (MAP) (Kaminski, 1986), are gradually being accepted throughout industry. The cost currently required for special software and hardware to translate protocols while messages are passed from one machine to another will be reduced as more automation suppliers adopt the MAP communications standards. Also, software systems designed to control and manage a wide variety of machines which communicate in different languages are gradually being commercialized (e.g., the TRANSCELL system described in Bourne and Fox, 1984). With this type of software available "off-the-shelf," neither users nor vendors will have to go through the expensive process of developing a customized version of a communication and control system for each sensor-based robotic system installed. The other major cost that could be reduced is the development cost for end-of-arm tooling and part-orientation hardware (which, for Level I robots, is assumed to be three times the base price of the $60,000 robot and twice the base price of the $100,000 robot) through improved ability to acquire unoriented parts and to find parts that are not precisely located.

Suppose a vision system could be purchased which has a greater capability (in terms of being programmable and able to transmit information) than the leading commercial systems offered in the early 1980s. Say the total cost of this hypothetical vision system, including all necessary software and its own computer, is $100,000. A key feature of this system is that, by making some simple alterations to the communication and control systems within the robot, the vision system and the robot are "plug-compatible." If the vision/robot system were able, within certain limits, to acquire randomly oriented parts and to locate parts whose positions are not precisely known, then the cost of designing and installing tooling and accessory hardware would be substantially reduced. Suppose that the development cost were roughly the same as the robot base price. The price of this hypothetical Level II installation would be as follows:

	Medium-Cost Robot	High-Cost Robot
Robot	$60,000	$100,000
Development cost	$60,000	$100,000
Vision system	$100,000	$100,000
Total	$220,000	$300,000

Given these assumptions, the cost of installing a Level II system would be essentially the same as installing a Level I system. If only one installation of a given application were installed, the payback period would be the same as for the Level I case examined earlier in Table 2-24. An important difference here is that robot users indicate there are perhaps three times as many potential applications for Level II systems as for Level I systems. There would be more potential applications per establishment, and even medium-sized and small establishments would have use for at least several robots. Suppose it were still the case that when more than one robot is installed, the development cost decreases with each successive installation for similar applications. Since there are more applications within each establishment, the average cost per installation would decrease and payback periods would be shorter than is indicated in Tables 2-28, 2-29, and 2-30, especially for small establishments.

In the Level I case discussed earlier, it was concluded that if the decision to install robots were based primarily on payback periods calculated from labor savings, only the largest establishments would have a clear-cut economic incentive to robotize. It appears that if Level II systems could be purchased and installed for approximately the same cost as Level I systems (substituting the cost of sensing for the cost of accessory tooling and parts-orientation hardware), the payback periods would decrease for most small establishments. If payback periods were sufficiently short, establishments with fewer than 1000 workers would have a strong economic incentive to install robots. If this were to happen, then there would be good reason to reconsider market forecasts which predict that there will be *at most* 150,000 robots in use in 1990.

There is even the possibility that vision and other types of sensing systems would eventually make Level II robots substantially cheaper to install than Level I systems. Clearly, if this were the case, payback periods would drop substantially, even without considering the decrease in the average cost per installation for multiple installations.

2.10. Displacement Impacts of Retrofitting Level I and Level II Robots to Their Full Potential

What would happen to labor requirements and production cost in a particular firm if Level I and Level II robots were used to perform all *the jobs that they are capable of doing?* First, a simple comparison is made between the number of workers that could be displaced by robots and the number of job openings that would occur if the size of the work force were to decrease as a result of attrition. The point of interest is the number of years it would take for the job openings resulting from attrition to equal the number of workers that could be displaced by Level I and Level II robots. Next, a simple model of a firm's production function is used to calculate the increase in the level of output that would be required to maintain the base level of production worker employment, even after robots are installed. For various assumptions regarding the percentage of production workers displaced by robots, the attrition rate, and the impact on factory throughput, the increase in the level of output that would be required to reabsorb all displaced workers is calculated.

2.10.1 Workers Displaced by Robots versus Job Openings Due to Attrition

The attrition rate is a measure of the percentage of workers who leave the work establishment as a result of quits, discharges, permanent disability, death, retirement, or transfers to other companies. Several studies have argued that in the foreseeable future, the displacement effects of robots can be nearly or completely offset by natural reductions in employment levels that occur as a result of attrition. For example, earlier in this chapter, Hunt and Hunt's (1983) displacement rate estimates were compared with Bureau of Labor Statistics projections of annual replacement needs to offset attrition (Table 2-9). Hunt and Hunt estimate that the number of workers displaced by robots per year will be fewer than what the BLS estimates will be required just to replace workers who leave their workplaces through attrition. If one were also to consider projections of workforce needs to meet increases in demand, as well as replacement, the BLS estimated rate of work-force increase would be even larger. Hunt and Hunt conclude that job displacement levels produced by robots in this decade will be low even when compared with replacement requirements of the labor force.

Krause (1982) notes that even if General Motors were to increase its robot population from 302 to 14,000 between 1980 and 1990, the 28,000 or so workers displaced would be more than offset by the 97,000 jobs vacated through voluntary turnover as a result of the 4.1 percent annual attrition rate of hourly employees. Based on an analysis of technological change in manufacturing industries from years 1957 through 1975, Jacobson and Levy (1983: 79) conclude that "employment reductions can be handled largely through attrition." Respondents to the 1985 Delphi survey were asked for opinions on what would happen to workers who are displaced by robots. The majority of respondents replied that between 0 and 5 percent of those displaced would end up being fired, that the vast majority would be transferred within the same plant or to a sister plant, and that the remainder would retire early or quit (Smith and Heytler, 1985: 181).

Estimates of the technical potential for replacing production workers with Level I and Level II robots were given earlier in the chapter in Table 2-13. The result was that about 12 percent of production worker jobs could potentially be replaced by Level I robots, and about 33 percent by Level II robots. Suppose this proportion of workers were to be replaced throughout a large company or an industry. How long a period would it take for the number of job vacancies cre-

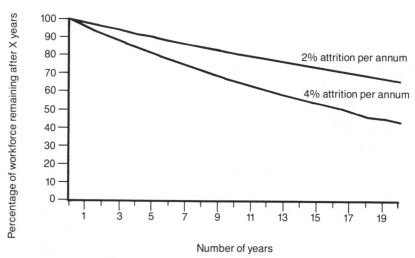

Figure 2-8: Decline in Size of a Workforce over Time if Job Vacancies Created by Attrition
Are Not Refilled

ated through attrition to equal the number of workers displaced by robot use?
Each curve in Figure 2-8 shows how the size of a work force would decline year
by year, assuming a given percentage of workers leave the firm annually and are
not replaced. The impacts of external market forces, such as the level of overall
economic activity, on demand levels and on work force size are ignored. Across
the durable goods industries, in which industries SIC 34–37 are included, attrition rates have averaged between 2 and 4 percent per annum, and these values
are used for the top and bottom curves in the Figure. For 12 percent of the jobs to
turn over, it would take seven years with the 2 percent attrition rate, and only
three years with the 4 percent rate. For a third of the jobs in the work force to
turnover, it would take nearly 20 and 10 years, respectively, with the low and
high attrition rates.

   In actuality, these figures should be viewed only as lower bounds on the
number of years it would take for job openings resulting from attrition to equal
workers displaced by robots, since at the industry level a given percentage of
displacement is not necessarily offset by the corresponding percentage of attrition rate. Presumably, this would also be true across a large company. This point
is more fully explained by Jacobson and Levy (1983: 77).

   A 5 percent employment decline cannot be fully offset by a 5 percent attrition rate.
There will be some displacements because the employment change across firms is not
uniform. Even while total employment in the industry is declining, some firms will be
expanding. Attrition in the firms that are expanding must be replaced by hiring and
obviously cannot count against the net decline. The dispersion of firms around the mean

employment change is substantial so that even where employment in the industry is constant about 1 percent of the industry's labor force will be undergoing displacement.

Even if the full potential for Level I robot use *were* to be realized over a 5–10-year period, the number of workers displaced would be substantially fewer than the number of job openings that would result from attrition. Few if any in the existing work force would involuntarily lose their employment. If the full potential for Level II robots were achieved within a decade, it appears there would in fact be a significant problem with displacement. However, for the reasons discussed in the analysis of the costs of installing robots, it is very unlikely that the full potential for Level II robots will be reached in such a short time. And if it were to take several decades to implement Level II robots to their full potential, it would also be the case that job displacement could be offset by job replacement needs from attrition.

2.10.2. Considering Displacement with Attrition and Moderate Increases in the Rate of Throughput

In the analysis that follows, various assumptions are made as to how the use of robots would decrease production labor requirements and alter the throughput of a factory. Two scenarios are considered. In the first, robots displace direct labor but have no effect on the factory's throughput. In the second, labor is displaced and the throughput of the factory is also increased by 20 percent. For each scenario, two cases are considered. In one case, the effects of attrition are not taken into account. In the other case, the work force is assumed to decrease in size because of attrition during the multiyear period in which workers are displaced. For each scenario, I have calculated the increase necessary to maintain production labor requirements at a level that would allow the firm to reabsorb all workers displaced by robots.

The discussion in the previous section considered the case where robot use has no impact on the factory's rate of throughput, and where there is attrition. Even if 12 percent of the workers were displaced over a period of five years or more, job vacancies due to attrition would be equal to or greater than the number of workers displaced. The added consideration here is the effect that robot use has on the throughput of the factory. In this case, the decrease in unit labor requirements is large enough so that attrition does not offset displacement. The question is how large an increase in the level of output would be required to maintain the base employment level. If the required increase were moderate, one might argue that employment levels could still be maintained. For example, productivity improvements resulting from the combination of decreased labor cost and increased throughput would lower prices. If demand were price sensitive, the resulting increase in demand could offset the effects of job displacement.

Before proceeding with the details, some of the assumptions underlying this framework should to be clarified in order to make the limitations of this analysis clear to the reader. First, issues relating to changes in skill requirements for a given occupation or changes in the overall occupation profile are not considered here. Therefore, it is implicitly assumed that if a production worker is displaced by a robot and if there is a need for an additional production worker resulting either from an increase in demand or from job turnover, then the displaced worker can be reabsorbed by the firm. That is, the skills required of production workers after the implementation of robots do not pose a barrier to reabsorbing the displaced workers. Second, new jobs that may be created as a result of using robots, such as systems programmers and robot maintainers, are not considered here.

Consider a cost function for a firm that is described by fixed technological input coefficients. In the neighborhood of the equilibrium level of output, it is assumed that the production function exhibits constant returns to scale, so that the price of the output equals average cost. Invoking the standard assumption of perfect competition in the marketplace, the value of the product is exactly equal to the sum of the costs of the factor inputs, and is given by

$$p_0 Y_0 = w_L L_0 + w_S S_0 + w_K K_0 + w_M M_0$$

where

p_0 = price of output in equilibrium before robots are used (base period).
Y_0 = quantity of output produced
L_0 = input of production labor
S_0 = input of salaried labor
K_0 = input of capital
M_0 = input of materials
w_i = price of input factor i, where i = L (labor), S (salary), K (capital) and M (materials).

The input coefficients (a_L, a_S, a_K, a_M) express the quantity of each factor input required to produce a unit of output. The base period price, p_0, can be expressed as a function of the prices of the input factors and of the input coefficients.

$$p_0 = w_L a_{L,0} + w_S a_{S,0} + w_K a_{K,0} + w_M a_{M,0}$$

where $a_{L,0} = L_0/Y_0$, and with $a_{S,0}$, $a_{K,0}$, and $a_{M,0}$ are similarly defined.

In this framework, a change in the technology of production is represented by a change in the magnitudes of the input coefficients. Consider the production labor input coefficient in the base period, $a_{L,0} = L_0/Y_0$. If robots are substituted for production workers, total production labor requirements, L_0, will decrease by ΔL percent. Values for ΔL are negative, reflecting the drop in labor requirements. Also, the use of robots may make it possible to increase the overall throughput of the facility from Y_0 to $Y_0(1 + \Delta T)$ units, where ΔT is the increase in the rate of throughput. If production labor requirements were to decrease by

ΔL percent, and if throughput were to increase by ΔT percent, it would take only $L_0(1 + \Delta L)/(1 + \Delta T)$ units of labor to produce Y_0 units of output. In summary, the revised input coefficient for production labor, $a_{L,*}$, in the case where robots are used is given as follows:

$$a_{L,*} = \frac{Y_0(1 + \Delta L)}{L_0(1 + \Delta T)}$$

$$= a_{L,0}\frac{(1 + \Delta L)}{(1 + \Delta T)}$$

$$= a_{L,0}[1 - \frac{(\Delta T - \Delta L)}{(1 + \Delta T)}] = a_{L,0}(- \Delta a_L)$$

where $(\Delta T - \Delta L)/(1 + \Delta T) = \Delta a_L$
$\Delta a_L = (a_{L,*} - a_{L,0})/a_{L,0}$

The new unit labor input coefficient, $a_{L,*}$, is expressed as the difference between the old labor coefficient, $a_{L,0}$, and the percentage of change in the old labor coefficient, Δa_L. If the input coefficients for salaried labor (a_S), capital (a_K), and materials (a_M) were to change in a similar fashion, the new price for the industry's output could be expressed as

$$p_* = w_L a_{L,0}(1 - \Delta a_L) + w_S a_{S,0}(1 - \Delta a_S) + w_K a_{K,0}(1 - \Delta a_K) \qquad (2.6)$$
$$+ w_M a_{M,0}(1 - \Delta a_M) = p_0 - w_L \Delta a_L - w_S \Delta a_S - w_K \Delta a_K - w_M \Delta a_M$$

The change in price is given by

$$\Delta p = \frac{p_* - p_0}{p_0}$$

When the right-hand side of Equation (2.6) is divided by p_0, it can be expressed in terms of factor shares, since

$$\frac{w_L a_{L,0}}{p_0} = \frac{w_L L_0}{Y_0 p_0} = v_L$$

where v_L = the share of production labor in total cost, and v_S, v_K, v_M are similarly defined. The percentage of change in cost can be simplified to

$$\Delta p = - v_L \Delta a_L - v_S \Delta a_S - v_K \Delta a_K - v_M \Delta a_M \qquad (2.7)$$

The percentage of change in price is a weighted average of the percentage of change in the input coefficients, where the weights are given by the factor shares in the base period (before the change in technology). The percentage of change in each input-requirement coefficient depends on the percentage of change in the total requirements for the factor and in the percentage of increase in throughput. With the four input factors defined above, the percentage of change in price is a function of nine variables (the four factor shares, the percentage of change in total input requirements for the four factor inputs, and the percentage of change in throughput).

The firm's new requirements for production labor are given by the product of the new input coefficient and the new level of output:

$$L_* = a_{L,0}(1 - \Delta a_L)\, Y_0(1 + \Delta Y)$$

Since the base level of employment, L_0, is equal to the product, $a_{L,0}Y_0$, the new requirement for production labor is

$$L_* = L_0(1 - \Delta a_L)\,(1 + \Delta Y) \qquad\qquad (2.8)$$

If the effects of attrition on the size of the work force are not considered, then the level of production worker employment required after the retrofit of robotics, L_*, to equal or exceed the base level of employment, L_0, is

$$L_0(1 - \Delta a_L)\,(1 + \Delta Y) \geq L_0 \qquad\qquad (2.9)$$

If the time it takes to implement the robotic technology and displace ΔL percent of the production workers were to span a multiyear period, Δx percent of the workers would permanently leave their place of work as a result of attrition. During this period, the size of the work force would decrease from L_0 to L_0 $(1 - \Delta x)$. The new total labor requirements would still be given by Equation 2.8. However, since Δx percent of the work force would have left, the condition that the new labor requirements equal or exceed the number of people still employed would be modified to

$$L_0(1 - \Delta a_L)\,(1 + \Delta Y) \geq (1 - \Delta x)L_0 \qquad\qquad (2.10)$$

What is a reasonable range of values for Δx? Assuming an attrition rate of r percent per annum, after n years, a work force of size L_0 decreases to a size of $(1 + r)^{-n} L_0$. For this discussion, it is assumed that Δx, the fraction of the workers that leave the workplace as a result of attrition during the time period considered, is 15 percent. With a 2 percent attrition rate, it would take eight years for the work force to decrease by this proportion. With a 4 percent rate, it would take about four years.

In the first scenario, robot use has no impact on the factory's throughput. It is assumed that total production labor requirements decrease by ΔL percent, and that the total requirements for the other factor inputs remain at their current levels. The parameters for Scenario 1 are given below.

Scenario 1: No Impact on Rate of Throughput

$\Delta T = 0$

$\Delta S = \Delta K = \Delta M = 0$

$\Delta a_s = \Delta a_K = \Delta a_M = 0$

$\Delta p = -v_L \Delta a_L - v_S \Delta a_S = v_K \Delta a_K - v_M \Delta a_M$ reduces to $\Delta p = -v_L \Delta a_L$.

$\Delta a_L = \dfrac{\Delta T - \Delta L}{1 + \Delta T}$ reduces to $\Delta a_L = -\Delta L$.

The level of output for which $L_* = L_0$ reduces to

$$\Delta Y = \dfrac{-\Delta L - \Delta x}{\Delta L(1 + \Delta L)} \qquad\qquad (2.11)$$

for $|\Delta L| > |\Delta x|$.

Equation 2.11 is solved using the following parameter values: $\Delta L = -5\%$, -10%, -20%, -30%, and $\Delta x = 0, 15\%$.

The second scenario differs from the first in only one important respect. It is assumed that the installation of robots results in a 20 percent increase in throughput as well as a decrease in production labor requirements. As in the first scenario, it is assumed that production labor requirements decrease by ΔL, and that requirements for salaried workers, capital inputs, and materials are held constant. Requirements for materials are also assumed to be constant, despite the increase in rate of throughput. A rationale for keeping material requirements constant is to assume that scrap requirements are reduced as a result of more consistent processing. As a result of assuming an increase in throughput, the input coefficients for salaried labor, capital, and materials decrease, and input coefficient for production labor decreases by an amount greater than if there is a decrease only in production labor requirements. The parameters for Scenario 2 are as follows:

Scenario 2: Moderate Increase in Rate of Throughput

$\Delta T = 0.2$

$\Delta S = \Delta K = \Delta M = 0$

$\Delta a_S = \Delta a_K = \Delta a_M = \dfrac{0.2 - 0}{1 + 0.2} = 0.167$

$\Delta p = -v_L \Delta a_L - v_S \Delta a_S - v_K \Delta a_K - v_M \Delta a_M$ reduces to

$\Delta p = -v_L \Delta a_L - [1 - v_L(0.167)]$ where $v_L = 10\%, 20\%, 30\%, 40\%$. (2.12)

Notes: $v_S + v_K + v_M = 1 - v_L$

$\Delta a_L = \dfrac{\Delta T - \Delta L}{1 + \Delta T}$ where $\Delta L = -5\%, -10\%, -20\%, -30\%$

For each scenario, the impact of robot use on the price of output is also calculated. The percentage of decrease in price depends on the percentage of production labor displaced and on the production-labor share of total costs. All aspects of cost changes are not considered. Increases in capital cost required to install robots are ignored in this analysis so that calculated price decreases can be viewed only as upper bounds. Only one firm (or industry) is considered here and interindustry transactions are ignored. An important characteristic of most metalworking industries is that they sell most of their output to other industries (especially to other metalworking industries) to be used as capital or material inputs. These interactions would result in larger price decreases than are indicated in the analysis. The percentage of price reduction for each scenario is discussed in Chapter 4, which focuses on the impact of robot use on production cost.

Table 2-31 shows the percentage of increase in output, ΔY, required to main-

Table 2-31: Percentage of Increase in Level of Output Required to Reabsorb Displaced Production Workers

Reduction in Production Labor Requirements (ΔL)	Scenario 1[a]		Scenario 2[b]	
	Without Attrition[c]	With Attrition[d]	Without Attrition[c]	With Attrition[d]
5%	5	–[e]	26	7
10%	11	–[e]	33	13
20%	25	6	50	27
30%	43	21	71	46

[a]No increase in throughput ($\Delta T = 0$).
[b]20 percent increase in throughput ($\Delta T = 0.2$).
[c]No job openings as a result of attrition during transition period ($\Delta x = 0$).
[d]Size of the work force declines by 15 percent as a result of attrition during the transition period ($\Delta x = 0.15$).
[e]The number of job openings due to attrition exceeds workers displaced by robots, so all displaced workers could be reabsorbed without an increase in the level of output.

tain the employment level, with and without the effects of attrition for both scenarios. The percentage of increase in output is derived from

$$\Delta Y = \frac{\Delta a_L - \Delta x}{1 - \Delta a_L}$$

The increase in throughput in Scenario 2 results in a much larger decrease in the unit labor requirement than in Scenario 1. Correspondingly, a much larger increase in output is needed in Scenario 2 to maintain base level employment requirements. For each scenario, the decrease in the size of the labor force that results when attrition is considered substantially reduces the size of the increase in demand needed to maintain base-level employment requirements.

First, attention is focused on the cases where the reduction in total production labor requirements is 10 percent or less ($\Delta L = -5$ percent and -10 percent) and where the effects of attrition are considered. This represents the case where Level I robots are retrofitted to their full potential. In Scenario 1, job openings due to labor turnover would exceed workers displaced by robots, so displaced workers could clearly be reabsorbed into job openings within the firm. In Scenario 2, only a modest increase in output would be required to maintain employment levels ($\Delta Y \leq 13$). Given the corresponding price decrease, a price elasticity of less than unity would stimulate a sufficient increase in output to maintain base employment levels.[26] For example, where $\Delta L = -10$ percent and with the assumed attrition rate, the break-even value of the price elasticity will range from 0.76 to 0.62, depending on the production labor share of the total cost. In fact, with the conditions stated above, if the price elasticity were equal

26. The price elasticity of demand, $d\log\Delta Y/d\log\Delta p$, is the percentage of increase in output that would be realized for each 1 percent reduction in price. The reduction in price, Δp, is given in tables in Chapter 4.

to unity, there would even be a need to hire additional workers under both scenarios. The conclusion derived for these cases is that *all* of the workers displaced by robots could reasonably be reabsorbed within the firm through the combined effects of price-induced increases in output and job openings due to labor turnover.

Attrition always occurs, so why consider the "without attrition" cases? There are situations where workers displaced by robots cannot be reabsorbed through transfers to job openings created by attrition. Older product lines may be discontinued, and all of the replacement-job openings may be filled by workers from these discontinued lines. Other technologies which displace workers may also being introduced within the factory. Perhaps workers displaced by these other technologies will fill the replacement openings. In addition, simplification and rationalization of designs to make products easier to manufacture may have major impacts on reducing labor requirements. Also, changes in managerial policies and practices, such as the growing trend to eliminate inspectors and have operators "self-inspect" their work, may also result in displacement. The point is that, for a variety of reasons, job openings may not be available to reabsorb the workers displaced by robots.

Without the availability of job openings due to attrition, it is unly that a 10 percent reduction in labor requirements could be offset by an increase in the level of output. In Scenario 2, without attrition, output levels would have to increase by nearly a third to maintain the base level of employment. In Scenario 1, the output level would have to increase by 11 percent. But the price reduction would be so small in this case (about 2 percent) that it is unlikely that price-induced increases in demand would stimulate output to the required level. In conclusion, it is not realistic to expect that price-induced increases in the level of output can offset the effects of robotic job displacement as some authors have suggested (e.g., Vedder [1982]).

The cases in Table 2-31 where 20 and 30 percent of production labor is displaced represent situations where Level II robots are retrofitted to their full potential. Again, the general conclusion is that if Level II robots were to be implemented to their full potential, displacement would result in job loss. Even with attrition, the level of output would have to increase by very large amounts, from one-quarter to one-half, to maintain the base level of employment in Scenario 2. The required output increases are much smaller in Scenario 1, but so are the corresponding price reductions. Again, it is not clear if price-induced increases in demand would stimulate output to the required level.

In summary, a highly simplified analysis is presented which indicates that Level I robots could be implemented to their full potential within a short period of time without causing job loss, as long as displaced workers could be reabsorbed into job openings created through attrition. It must be emphasized that this conclusion hinges on the assumption that workers displaced by robotics can and will be reabsorbed within their firm. If, for some reason, these job openings are not available to those displaced by robotics, then the conclusion will no longer apply. This point shows that there are many limitations in considering only the impacts of robotics without considering the broader set of technological changes occurring within the factory.

Aside from the limitations pointed out here, this analysis suggests that it is necessary to have a more thorough and complete understanding of the effects of robotics on labor requirements in order to analyze further the labor impact issues in industries using robots. It is important to know if the economic bene-fits of robot use are restricted to reductions in labor cost, or whether they might also include an increased rate of throughput. It is also very important to know about the rate of job turnover, since attrition is often cited as the primary way to offset the displacement effects of robots. More detailed information on attrition rates is needed to confirm the conclusion that Level I robots can be implemented to their full potential within the decade with attrition fully offsetting displace-ment. This analysis also suggests that there is a need to take a more detailed look at the rate of development of sensor-based robot systems, since the implementa-tion of Level II robots to their full potential over a decade or so would result in a level of job displacement beyond that which could be offset by attrition.

2.11. Perspective on Findings

The findings of the first group of studies analyzing the impacts of industrial robot use on job displacement are summarized in Table 2-32. Do the results of my analysis change the basic message of the earlier studies that preceded it? No, in fact, they provide additional evidence to reinforce the message (Table 2-33). The evidence supporting the hypothesis that robot use will continue to be con-centrated within large firms for the next several years also supports the point that displacement impacts will be concentrated within particular geographic loca-tions and industries. The analysis of displacement impacts of utilizing Level I robots to their full potential supports the contention that even where displace-ment impacts will be concentrated very few workers will actually lose their employment as a direct result of robot use. Overall, the findings from the initial studies, as well as from the additional work presented in my 1983 analysis and this update, support what has been referred to as the conventional economist's view of the impacts of technological change on the labor force. This view is summarized below, from a study of the impact of automation on the work force sponsored by the Office of Technology Assessment (OTA, 1983: 5).

Although new technologies may be created at varying rates, the conventional view among economists is that the use of new technologies spreads relatively slowly. It is commonly assumed that firms adopt new technology in a rational fashion, meaning that

Table 2-32: Summary of Initial Studies on Job Displacement Impacts of Industrial Robot Use

---

Eikonix Study (1979)
Time horizon:　　　　20 years (until year 2000).
Fraction of manufacturing production workers displaced in time horizon:
　　　　　　　　　15% by Level I robots,[a] 55% by Level II robots.[a]
General conclusions: There will be a slow but steady increase in robot population in the industrial sector.
　　　　　　　　　Robots will not cause massive numbers of people to lose their jobs. Local and
　　　　　　　　　structural employment effects will be noted, but these effects will be swamped by
　　　　　　　　　the overall state of the economy.
Ayres and Miller Studies (1982–1983)
Time horizon:　　　　One to three decades.
Fraction of manufacturing production workers displaced in time horizon:
　　　　　　　　　14% by Level I robots,[b] 40% by Level II robots.[b]
General conclusions: Robots will be able to replace significant numbers of machine and tool operators
　　　　　　　　　and unskilled laborers in the durable-goods industries (especially in the metal-
　　　　　　　　　working sector) over the next several decades. Displacement impacts are not the
　　　　　　　　　result of the total number of people affected but of the concentration of effects in a
　　　　　　　　　relatively narrow sectoral, occupational, and regional setting. Displacement will
　　　　　　　　　be of sufficient magnitude in some industries and regions, and for some groups of
　　　　　　　　　workers, to be a cause of concern.
Hunt and Hunt Study (1983)
Time horizon:　　　　Seven years (until 1990).
Fraction of manufacturing production workers displaced in time horizon:
　　　　　　　　　1–2% by Level I robots.[c]
General conclusions: Number of robots will increase less spectacularly than had first been anticipated.
　　　　　　　　　While there will be no general job displacement problem, there will be particular
　　　　　　　　　pockets of displacement that may cause labor-market distress. Particular occupa-
　　　　　　　　　tions, industries, and locations will suffer the brunt of the job displacement impact.
　　　　　　　　　There will be few workers actually thrown out of work even in highly impacted
　　　　　　　　　areas. The impact of displacement will primarily be to eliminate job openings.

---

　[a]Estimated percentage of operative workers whose jobs could be performed by robots by the year
2000.
　[b]Estimated percentage of selected metalworking craftworkers, operative workers, and laborers whose
jobs could be performed by robots over a horizon of two to three decades.
　[c]Estimated percentage of operative workers and laborers displaced by robots in 1990, given stated
assumptions regarding the size of the robot population in the United States.

Table 2-33: Summary of Miller Study on Job Displacement Impacts of Industrial Robot Use

---

Miller Study (1983 and this revision thereof)
Time horizon:　　　　1–3 decades.
Fraction of manufacturing production workers displaced in time horizon:
　　　　　　　　　12% by Level I robots,[a] 33% by Level II robots.[a]
General conclusions: Large firms have a substantial cost advantage over small firms in using robots. As
　　　　　　　　　a result, over the next decade or so, robot use will be primarily concentrated within
　　　　　　　　　large firms. If the estimated potential for Level I robot use were fully realized, but
　　　　　　　　　only within large manufacturing establishments, the implied number of robots in
　　　　　　　　　use would approximate the market forecast for the cumulative U.S. robot popula-
　　　　　　　　　tion in the mid-1990s. Even if the full potential for Level I robot use were realized
　　　　　　　　　over the next 5–10 years, it is reasonable to expect that all displaced workers
　　　　　　　　　would be reabsorbed within the firm through the combined effect of job openings
　　　　　　　　　due to attrition and price-induced increases in output.

---

　[a]Estimated percentage of all production workers in manufacturing (craftworkers, operative workers,
and laborers) whose jobs could be performed by robots over a horizon of one to three decades.

they strive to use the most affordable processes to avoid the cost of prematurely scrapping facilities and to adapt technologies to their individual needs. This view implies that, since firms typically do not adopt each technological advance as soon as it is developed and since firms experience some normal level of employee turnover, employees are not (repeatedly) subject to catastrophic displacement.

How believable or trustworthy are the findings of the studies in Tables 2-32 and 2-33? Also, how can the limitations of studies analyzing the impacts of robot use on labor-force requirements be put in perspective so they are clearly understood? These are major points of concern, since analysis of the general impact of technological change and of the particular impact of automation on the labor force is generally acknowledged to be a difficult and elusive task. For example, Hunt and Hunt (1985: 7–10) identify six "dangers endemic to studying the subject" of the impact of technological change on the labor force.

1. The revolutionary aspects of any new technology tend to be exaggerated.
2. Assumptions regarding the diffusion of new technologies tend to be excessively optimistic.
3. It is far more difficult to identify the new jobs that will be created in the future by new technology than to identify the jobs that will likely be lost.
4. There tends to be inappropriate generalization of results from particular case studies.
5. Available data and statistics tend to be misused and misunderstood.
6. In general, this type of forecasting strains the limits of what is truly knowable in advance of events.

Hunt and Hunt (1985: 2) observe that "studying the employment effects of technological changes with currently available data is like trying to put together a jigsaw puzzle with many of the pieces missing—most of the pieces that are present do not fit together." The Office of Technology Assessment (OTA, 1986: 362) similarly comments, "Given the difficulties inherent in predicting technological innovation and diffusion, projections of the effects of technology on jobs, or even of the technologies themselves, have many uncertainties."

In its report on the impact of factory automation, the Office of Technology Assessment (OTA, 1984: 104) explains why the "analysis of the employment impacts of programmable automation [PA] is fraught with difficulties."

Briefly, analysts generally approach the problem from two perspectives: the engineering approach, which focuses on the potential for equipment to substitute for people on a task-by-task basis; and the economic approach, which derives employment estimates from models of the interaction among industries based on their requirements for labor and other production inputs. Both approaches have shortcomings. Moreover, the number of different PA technologies, the range of equipment designs and implementation strategies, and uncertainties about the speed and success of technical advances make the formulation of general rules about job loss (or creation) risky. So, too, does existing variation among employers (even in the same industry) in job mix, job definition, and adaptability to change. Finally, data describing prevailing skill requirements, jobs, and job mixes among firms are limited.

The OTA study makes an important distinction between two contrasting approaches typically used to analyze the impacts of automation on the labor force: the *engineering* approach and the *economics* approach. In using the engineering approach, estimates of the impact of automation on the labor force "are made by describing the capabilities (for physical and mental work) of new automation technologies, projecting capability improvement over time, comparing the capabilities to tasks performed by humans, relating human tasks to different occupations, and deriving the number of jobs, by occupation, that could be assumed by new and future improved types of equipment. This is done by comparing guesses as to the percentage of work that could be transferred to programmable automation with counts of the number of people currently doing that work" (OTA, 1983: 14).

The earlier Ayres and Miller (1982, 1983) estimates of potential displacement, as well as my revised estimates presented in this chapter, were clearly derived using what is described above as the engineering approach. Yet previous studies have identified three general shortcomings in this approach (OTA, 1983: 83). First, there are always uncertainties regarding the capabilities of new automation. For example, it is difficult to comprehend the full potential for robots or other forms of programmable automation to perform jobs in ways other than by stimulating human behavior, or to perform jobs that are poorly done or not done at all by humans because of human limitations. This lack of foresight may lead to an over- or underestimation of job displacement. Also, there will inevitably be errors in projecting future technological capabilities of robots. One might argue that in my analysis, this particular shortcoming is mitigated to some extent by two factors. The first factor is that estimates of the fraction of jobs robots can perform are taken from a survey of experienced robot users, and these people should know about the potential for using robots in their own production environments. Also, more recent surveys of other robot users have produced similar results, indicating that people's opinions on this subject are reliable. The second factor is that the discussion emphasizes the impacts of using Level I robots, the capabilities of which are well known, since they have been available for well over a decade. Thus, for the most part, the analysis is not contingent upon accurate forecasts of the emerging capabilities of more sophisticated types of robots. Yet, despite these mitigating factors, it must be acknowledged that there is still uncertainty about both the current and future potential for implementing robotic technology. Just because the survey results seem to be consistent does not mean they are correct!

The second shortcoming of the engineering approach is that it does not systematically consider the complexities of implementing the technology within a particular organization. For example, the number of people employed to work with an automated system may reflect complex management and implementation considerations that are independent of the capabilities of the specific type of equipment. For example, a risk-averse manager may provide redundant capabilities in the form of extra or overskilled workers for manual-performance backup or monitoring services, at least in the short term when programmable automation is relatively unfamiliar. It is in this sense that engineering-based estimates of displacement may be misleading, because they are derived from a technically ideal mix of humans and equipment, whereas in practice this ideal mix may not be used. Because of the ways managers might actually choose to implement robotic technology, there could be a substantial difference between the number of jobs that are robotized and the number of workers who will be displaced from their jobs. This shortcoming could be offset to some extent by varying the assumptions about the mix of humans and machines that would be used in practice.

The third shortcoming of the engineering approach is that it ignores the influences of macroeconomic factors on labor requirements. As pointed out in OTA, 1983 (p. 15), the consequences of job displacement and creation will depend not only on how robotics and other forms of programmable automation affect the number and type of tasks per worker, but also on changes in sales volume and mix of products—which together finally determine the total number of tasks done. The overall level of economic activity and the composition of goods and services required may vary in response to factors other than technological change, such as shifts in consumer tastes. In addition, employment consequences of programmable automation will depend on the number and type of people willing and able to work at different types of jobs, which also can vary independent of technology.

The OTA studies point out that the engineering approach to analyzing labor-force impacts is useful for identifying the types of people who may be affected by robotics and other forms of automation in the workplace. However, as currently used, this approach is often too simplistic to provide realistic estimates of industry or economywide employment change (OTA, 1983: 15). The use of large-scale economic models is required to systematically evaluate whether persons displaced from particular industries may find other job opportunities requiring their skills, and therefore whether job displacement is likely to lead to unemployment. Estimates of labor-force impacts based on economywide models—which take into account the growth and decline of different industries, the likelihood of whether individual industries adopting new technologies will maintain or increase output levels, and the responsiveness of industry employment levels to technological change—prevent overattributing employment changes to single influences such as technological change. The key point is that macroeconomic models show the consequences of combinations of influences on labor-force requirements (OTA, 1983: 18). Estimating labor-force impacts through the use of macroeconomic models also has shortcomings. For example, as pointed out by the OTA (1983: 19), the use of large-scale models tends to oversimplify complex processes, and it is unclear how well these models account for changes in equipment technologies or capture the impacts of non-traditional types of equipment.

What conclusions would be reached if the economic approach were used to evaluate the impact of robotics on the labor force? The study by Leontief and Duchin (1984) is the most comprehensive analysis of this type which specifically addresses this topic. The study uses an input-output model of the U.S. economy to examine the potential effects of computer-based technologies on the growth and decline in the number of jobs by occupation and industry, based upon different sets of assumptions about the rates at which computer-based technologies will spread throughout the economy. The study specifically considers the labor-force impacts of the increased use of robots and computer-numerically controlled machine tools in manufacturing industries.

Leontief and Duchin show that increased use of robots and other forms of computerized automation in both the factory and the office will displace a substantial number of workers. Their analysis indicates that "the intensive use of automation will make it possible to achieve over the next 20 years significant economies in labor relative to the production of the same bills of goods with the mix of technologies currently in place" (Leontief and Duchin, 1984: 1.15). They estimate that with increased use of computers and computer-controlled machines, it will be possible to conserve about 10 percent of labor input relative to what would have been required to meet a comparable level of demand with current levels of technology. The study also indicates that the total number of jobs in the economy will continue to increase between 1980 and 2000. They estimate that even with the substantial increase in the use of computers and computer-controlled machines in the factory and office, total employment will still increase by tens of millions in the 20-year period between 1980 and 2000.

The Leontief and Duchin study shows that even in those production worker occupations where robot use is concentrated the number of workers employed is projected to increase between 1980 and 2000. The study assumes that robot use will displace production workers in the six occupations shown in Table 2-34.[27]

27. Robot use is also assumed to result in the displacement of inspectors. Leontief and Duchin (1984) project a decline in the number of inspectors required per each $1 million of output due to greater accuracy and dependability of more highly automated manufacturing processes. However, the existing literature on robot use is not sufficient for projecting displacement of inspectors by industry, so inspectors are not shown in the table. The future use of industrial robots within the six

Table 2-34: Estimated Number of Workers Displaced for Each $1 Million Worth of Robots Used, for Three Industrial Sectors

	Industry		
Occupation	Primary Iron and Steel	Household Appliances	Motor Vehicles
Assemblers	0	14.30	10.10
Packers and wrappers	1.45	1.43	0.72
Painters	0	1.78	4.33
Welders and flame cutters	3.63	3.57	7.58
Other machine operators	28.50	12.50	11.90
Laborers	2.18	2.14	1.08

Source: Leontief and Duchin, 1984: 4.42.
Average unit price of a robot assumed to be $84,000.

The figures for each occupation in the table show the number of workers that will be displaced per $1 million of investment in robots in selected industries. For example, their framework assumes that for every $1 million of investment in robots in the motor vehicle industry, nearly 40 workers will be displaced (12 semiskilled machine operators, 10 assemblers, 8 welders, etc.). Leontief and Duchin (1984: 1.15) estimate that the reduction in the demand for production workers which is directly attributable to robots will be about 400,000 in 1990 and almost 2 million in 2000, assuming rapid growth in robot use (i.e., a 25 percent increase in robot use per year between 1980 and 2000). Under this scenario, there will be 127,000 robots in use in 1990 and 316,000 in use in 2000. They also point out that the net demand for these same types of workers over this period will be about the same as will be realized in the baseline scenario (i.e., where there is no further spread in the use of robots throughout industry beyond the extent of their use in 1980). The authors explain that the greater need to produce higher levels of output in the capital goods industries under the first scenario (which assumes increased use of automation) offsets the displacement effects of robot use and keeps employment levels for semiskilled production workers at about the same level.

A more detailed look at employment projections for assemblers and welders, two production occupations that will be heavily impacted by robot use over the next 20 years, is given in Table 2-35. With increased use of robots, the demand

---

occupations considered is projected on the basis of two types of investment: robots purchased to modernize existing plants and those purchased for new plants. The dollar value of robots used per each $1 million of output per plant was assumed to differ between these two types of investment. It was also assumed that one robot would do the job of three workers over a two-shift operation. A concise summary of the assumptions used in the Leontief and Duchin study are given in Romero, Toye, and Baldwin, 1984.

Table 2-35: Comparison of Estimated Employment Levels for Welders and Assemblers, with and without Increased Robot Use, 1980–2000

	Estimated Employment With Increased Use of Robots[a]	Without Increased Use of Robots[b]	Percent Reduction in Employment Due to Robot Use[c]
		Assemblers	
1980	1,165,567	1,167,322	0.2
1985[d]	1,650,102	1,676,245	1.6
1990	1,531,340	1,629,108	6.0
1995	1,562,929	1,747,762	10.6
2000	1,680,043	1,965,400	14.5
		Welders	
1980	682,309	683,473	0.2
1985[d]	872,895	885,887	1.5
1990	775,181	819,005	5.3
1995	802,221	883,305	9.2
2000	849,635	970,352	12.4

Source: Faye Duchin, pers. comm., 1985.

[a]Scenario assumes that the average use of robots per unit of output will grow at a real rate of 25 percent per year from 1980 until 2000. Constraints on available labor, according to current labor-force projections, are considered in estimating attainable levels of final demand. See Scenario 4 in Leontief and Duchin, 1984.

[b]Scenario assumes no further automation or any other technological change after 1980. From 1980, robots are used only to the extent that they figured in the average technologies of each industry that prevailed in 1980. See Scenario 1 in Leontief and Duchin, 1984.

[c]Computed as (Column 2-Column 1)/Column 2.

[d]These numbers are large because the model shows 1985 to be at the top of an investment peak.

for assemblers can be expected to increase by 514,000 workers between 1980 and 2000. Without increased use of robots, the table results show that the demand for assemblers can be expected to increase by nearly 798,000 during this period. Even with increased robot use in assembly occupations, demand for assemblers would still increase by a substantial amount, however, the number of assemblers in the year 2000 would be nearly 14 percent less than it would have been without increased robot use. The employment projections for welders under the scenarios with and without increased robot use illustrate the same point. Between 1980 and 2000, the total number of welders can be expected to increase under both scenarios. With increased robot use, the number of welders in the year 2000 would be nearly 12 percent less than it would have been without increased robot use. Faye Duchin (pers. comm., 1985) has commented, "While the annual growth in the number of workers is confounded by the presence of an investment cycle which particularly affects production workers, there is a secular increase in the numbers of both assemblers and welders over the period despite the direct displacement of almost 300,000 assemblers and 120,000 welders in 2000."

In summary, based on an evaluation of the impact of robotics and other forms of automation on the labor force using an economic approach, Leontief and Duchin (1984) project that there will be a substantial number of new jobs created despite a substantial amount of job displacement. In the jobs where people will be displaced by robots (the semiskilled production worker occupations in manufacturing industries), it is still expected that requirements for people will increase over the next 20 years, even under the scenario assuming "high growth" of robot use. These findings basically support the message regarding the labor-force impact of robot use stated at the beginning of this section. Robot use will clearly not result in a mass displacement of workers and relatively few of the workers displaced by robots will actually lose their jobs, because most (if not all) displaced workers will be reabsorbed into the work force. Whereas the studies reviewed earlier (Table 2-32) argue that the displaced worker would be absorbed by a replacement-job opening created through attrition, the Leontief and Duchin analysis shows that many displaced workers will be reabsorbed into jobs created as a result of overall increase in levels of demand.

A question raised in Section 2.10 is, What would happen to labor-force requirements if other forms of technology were being implemented in the factory along with robots? The Leontief and Duchin study partly addresses this issue, since the study considers the labor-force implications of increasing the use of robots, computer-controlled machine tools, and computers in the factory. The results indicate that in the years 1990 and 2000 the relative displacement of skilled machinists and tool and die makers by numerically controlled machines would be far greater than the displacement of semiskilled production workers by robots. Still, absolute requirements for numbers of skilled machinists are projected to increase from 1980 to 1990 and 2000, even under the scenario for increased use of computer-numerically controlled machines. An important point is that workers in the factory will be displaced by many other types of technology in addition to robots. In the Section 2.10 analysis of whether displaced workers can be reabsorbed, it was noted that replacement jobs created through attrition may not be available to workers displaced by robots, because these jobs may be filled by workers displaced by other forms of technological change. The Leontief and Duchin analysis indicates that in fact a sizable number of workers will be seeking replacement or new job openings because of displacement by several types of technologies. This raises the issue that, if overall levels of economic activity in the future do not turn out to be as favorable as is currently projected, it would be unreasonable to assume that all workers displaced by robots could be reabsorbed by attrition, since in the absence of the creation of new jobs, workers displaced by other technologies would also be competing for these same replacement-job openings.

The OTA study on factory automation also considers the labor-force implications of increasing the usage of several types of technology in the factory simultaneously. Based on the results of the Leontief and Duchin (1984) study and on all available evidence found with both the engineering and economic approaches, it concludes that job displacement resulting from increased use of programmable automation in the factory will not result in an increase in unemployment. The report (OTA, 1984: 104) states:

> OTA shares the view that the use of programmable automation can grow, as is expected, without large increases in national unemployment during this century. The effects on employment of labor-saving technologies can be offset in the aggregate by changes in the labor force, as well as by likely increases in output for capital goods and other products. Such output growth, of course, assumes a strong economy.

As in the Leontief and Duchin study, the OTA study's message is that macroeconomic factors will offset reductions in unit labor requirements in the factory brought about by the use of labor-saving technology. It elaborates on why worker displacement due to automation will not result in job loss as long as there is economic growth.

> Technology used in production is a secondary influence that is dwarfed by the effects of demand changes; it governs the mix of labor and other inputs. Technology change generally affects employment much more slowly than do demand shifts, because it does not affect an industry or an economy all at once. Automation, in particular, is typically adopted during periods of economic expansion, a timing that facilitates the adjustment of work forces through attrition (OTA, 1984: 102–103).

The OTA (1986: 321) reemphasizes this point in a more recent report on worker displacement. "Although technology has significant effects on the availability of jobs, aggregate demand for goods and services remains the key."

A conclusion of the initial studies on the impact of robot use already mentioned in (Tables 2-32 and 2-33) is that the key concern regarding worker displacement is not the absolute number of people affected but their concentration in particular occupations, industries, and regions. The OTA study, which considers the impact of factory automation more generally, makes a similar point (1984: 5). "Programmable automation is not likely to generate significant net national unemployment in the near term, but its use may exacerbate regional unemployment problems, especially in the East North Central and Middle Atlantic areas where metalworking industries are concentrated."

Studies to date indicate that the increased use of robots in particular and factory automation in general will not cause a sudden reduction in the number of workers required, and that in the aggregate there will be more than enough job openings in the future to offset displacement. David (1982: 154) makes a cautionary comment on interpreting the results of macrolevel studies on the labor force impacts of technological change.

". . . The extent of plant level job displacement connected with productivity-enhancing technological change is almost certainly greater than one would be led to surmise from empirical studies carried out at the industry level, or still higher levels of aggregation. Over any era in which the demand for an industrys products has been expanding, jobs lost through the modernization of existing production facilities are likely to be offset in some part, if not entirely compensated for, by jobs created through the opening of new plants in different locations. The magnitude of the required labor market reallocations is therefore generally under-represented by statistical studies which find weak or non-existent job-displacement effects of technological change at the industry, sector, or domestic-economy levels."

David's comment on the need for displaced workers to relocate to find job openings illustrates that technologically induced displacement still raises important transitional issues, despite the conclusion that there will be a net increase in employment at the economywide level.

The major emerging transitional issue related to technologically induced displacement is not one of whether there will be jobs for people in manufacturing facilities that use increasingly higher levels of automation, but whether workers displaced from their current jobs will have the skills required for the new ones. This was first alluded to in the comment made by Hunt and Hunt (1983) that the most remarkable thing about robot use is the skill-twist that emerges so clearly when the jobs eliminated are compared with the jobs created. Many other authors have also highlighted the changing requirements for skills as the major impact of automation. An analysis of the impacts of microelectronics and automation on jobs conducted by OECD (1982: 113) concludes, "The changes brought about by microelectronics to 1990 will have a greater effect on skill levels and organization structure than on the aggregate demand for labor." The OTA factory-automation study (1984: 144) notes: ". . . The proportion of skills and occupations found in manufacturing will shift substantially because of programmable automation. In fact, to date this impact has been more striking than any change in the level of employment." Based on a case study of a vehicle-assembly plant that was modernized from a manual to a highly automated facility, Miller and Bereiter (1985) comment that the most striking result emerging from a comparison of the old and new work force is the change in the mix of labor required. The OTA study on displaced workers notes (1986: 321), "Computer-based technologies and manufacturing systems require skills that are qualitatively different from the skills that many displaced workers possess." Leontief and Duchin (1984: 1.33) come to a similar conclusion and point out that "a major consideration in realizing the transition from the old to the new technologies will be the availability of workers with the training and skills that match the work that needs to be done."

3

Distribution of Value Added
across Batch
and Mass Modes of Production

3.1. Overview

Most arguments citing the need to accelerate the development and use of robotics and other types of flexible production technology are founded on two widely accepted premises:

1. Products which are batch produced are much more expensive than products which are mass produced.
2. Most of the value added in the metalworking industries (SIC 34–37) is accounted for by batch production.

It is typically argued that much of the value added within the metalworking sector is a type of penalty cost which is unavoidable because of the inherent inefficiency of batch production relative to mass production. For example, one widely cited study—Cook, 1975—claims that for a typical machined product, the unit cost would be 100–500 times lower if the product were produced using the most efficient mass production techniques than if it were produced in a one-of-a-kind mode, and 10–30 times lower than if it were batch produced.

This chapter examines the two premises about the nature of discrete-parts manufacturing. It provides a foundation for the analysis of production costs in the following chapter. Chapter 4 examines the effect on production costs the use of robotics technology could have if it were to diminish some of the differences between facilities organized to mass produce standardized products and those organized to batch produce customized products. Chapter 3 begins with a review of the microeconomic relationships between the volume of output produced, the characteristics of technology used for production, and the unit cost of production at the plant level. Then, the relationship between the level of unit cost and the level of output produced is examined across 101 different metalworking industries. Estimates of purchases of basic and processed metals are used as measures of the level of output of each industry.[1] Value added per unit and units of output are computed for each industry using pounds of metal processed as the standardized unit of output.

The basic structural relationships that underlie the shape of a "neoclassical" long-run unit cost curve for a particular product are also apparent in the comparison of unit cost versus output *across* industries. Among the 101 metalworking industries analyzed, the following relationships can be found:

1. Capital costs for equipment and machinery per unit of output decrease as the units of output produced increase.
2. Production labor costs per unit of output decrease as the units of output produced increase.
3. Value added per unit of output decreases as the units of output produced increase.
4. Machine utilization increases as units of output produced increase.

The implication is that the trade-offs which most strongly affect the organization of production within a particular plant—organized to make either small volumes of specialized products at a high cost or organizing to make large volumes of standardized products at a low cost—are also affecting the organization of production *across* industries. It is therefore argued that the dominant mode of technology used within an industry can be inferred from the industry's measures of pounds of metal processed and unit cost. Industries with the highest production costs per pound of output and with the fewest pounds of output are classified as custom and small-batch producers. Industries with the lowest production costs per pound of output and with the most pounds of output are classified as mass producers. The remaining industries, those with medium-range levels both of production costs per pound of output and of pounds of output, are classified as batch producers. With these assumptions, the proportion of value added and of output accounted for by metalworking products which are custom, batch, and mass produced is estimated in this chapter. The result of this analysis is that the industries with the highest levels of output and with the lowest levels

---

1. The term *basic metals* refers to inputs of "raw" metal stock: steel, brass, and aluminum in the form of bars, billets, sheets, strips, plates, pipe, tubes, etc., as well as castings and forgings made of the three basic metals. The term *processed metals* refers to inputs which are themselves the products of the industries in major groups SIC 34–38. In general, these products are basic metals which have been further processed within the metalworking industry.

of unit cost (which are assumed to be the mass producers) account *at most* for 25 percent of the value added and for less than 35 percent of the total output of the 101 industries in the sample. This analysis corroborates the frequently made assertion that 50–75 percent of the value added in the metalworking sector is accounted for by products which are batch produced.

3.2 Unit Cost Versus the Volume of Output

A central issue in this analysis is the relationship between the number of identical or similar units produced within a factory and the average cost per unit. Cook (1975) estimates that the cost of machining typical industrial parts declines by a factor of 100 or more in going from one-of-a-kind production to the most efficient mass production. He cites the example of producing a V-8 automobile engine block.

Under mass-production conditions, where the engine block is conveyed automatically along a transfer line, with the various operations (drilling, tapping, boring, milling and so on) being executed in sequence at the different stations along the line, the complete machining cost (excluding the raw material but including labor, tooling, machine depreciation and interest on investment) would be of the order of $25. If, however, only a few special cylinder blocks were to be made with general-purpose machines and skilled labor, the machining cost per block could easily rise from $25 to $2,500 and more. (p. 25)

Cook also estimates that those industries which batch produce metal parts are characterized by unit costs from 10 to 30 times higher than those in the mass-production sector of the economy. His rough estimates of the relative average cost per unit as a function of the number of units produced are shown in Figure 3.1. According to Cook's data, for typical industrial products which are machined, unit cost falls by a factor of 100-500 as volume increases six orders of magnitude from 1 to 1 million.[2]

Why does unit cost decrease in a regular fashion over the entire range of volumes? The fact that unit cost decreases as volume increases is not simply a matter of increasing the level of output in a particular plant (with a particular type of technology). Implicit in any cost curve spanning such a range of volumes is the assumption that production is organized in an optimal, cost minimizing way in each of the custom-, batch-, and mass-production regimes. Since all

2. Aside from the example of the V-8 automobile engine block, Cook offers no empirical evidence to verify his estimates of this unit cost curve for typical machined products. However, over a several-year period, he has simulated the cost of producing parts using various types of manufacturing systems (Cook et al., 1978).

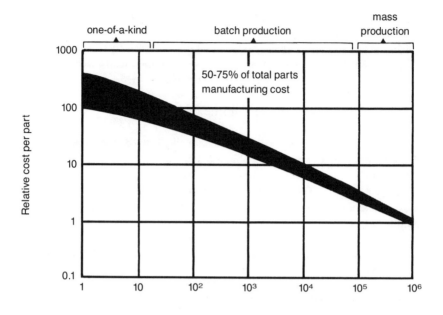

Production rate (parts per year)

Figure 3-1: Relationship between Cost per Unit and Number of Units Produced (*Source:* Cook, 1975)

possible technologies are considered in the composition of Figure 3-1, the curve represents the cost-minimizing *envelope* of production cost, or the long-run cost curve.

That average unit cost decreases as the level of output increases is a fundamental tenet of microeconomic-production theory which applies to all types of technology. Until capacity constraints are reached, increasing the level of output spreads the fixed-cost component over a larger number of units. Assuming that, for a particular production technology, variable factor requirements are nearly proportional to the level of output, average variable costs are nearly constant. Since average fixed costs decrease and average variable costs are nearly constant, average cost decreases as the level of output increases. In the short run, when the capacity of fixed resources, such as capital equipment or floor space, is fixed, it is typically argued that once output exceeds practical capacity, average cost per unit begins to increase. If capacity constraints are exceeded, the marginal cost of producing an additional unit exceeds the average cost of producing the previous unit. In other words, average cost per unit increases when output levels increase beyond preferred capacity levels. This is the rationale behind classical U-shaped average cost curves.[3] If all input quantities

3. See Varian, 1978, pages 22–27, for an overview of classical U-shaped cost curves.

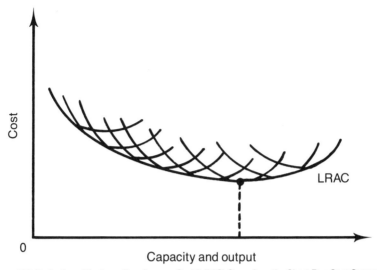

Figure 3-2: Derivation of the Long-Run Average Cost (LRAC) Curve from the Short-Run Cost Curves, for a Family of Technologies (after Rosegger, 1980, Figure 4.2)

can be varied, including those which are fixed in the short run, the producer can alter the technology and, up to some limit, achieve lower unit cost as volume increases. Figure 3-2 shows the way in which the long-run cost curve for a family of technologies is derived from the envelope of the U-shaped short-run cost curves. It is typically assumed that even the long-run average cost curve begins to increase at some very large level of output, because production gets so large that there are diseconomies of scale. This point is acknowledged here but will not be mentioned again.

For purposes of discussion, suppose that there are only three types of technology available, and that each technology has its own niche in the low-, medium-, and high-volume range. Examples of cost curves for technologies typically used for custom, batch, and mass production are shown in Figure 3-3.[4] The curve labeled "piece production" represents a labor-intensive technology which is typical of custom production. Fixed-capital requirements are lowest

4. These cost curves are not U-shaped, since it is assumed that additional plants using the same technology could be replicated at the point where demand exceeded existing capacity. The "step" increases in cost which would occur each time a plant is replicated are not shown. Other examples of cost curves for custom-, batch-, and mass-production technologies, derived from detailed engineering estimates of machine and labor requirements are given in Abraham, Csakvary, and Boothroyd, 1977, Whitney et al., 1981, and Boothroyd, Poli, and Murch, 1982. These studies focus on the assembly of various metalworking products.

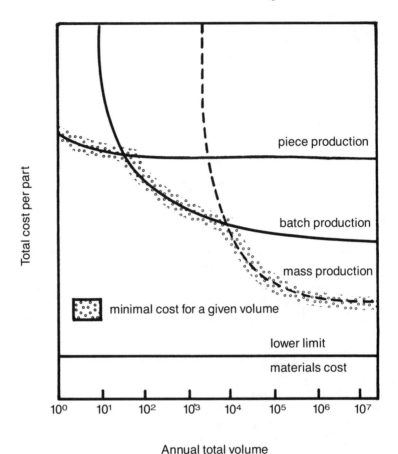

Figure 3-3: Average Cost versus Batch Size is Shown by a Simplified Curve Resulting from a Complex Curve that Includes the "Step" Changes of Adding New Machines Which Are Not Fully Utilized to the Overall Cost Factor of the Product (after Borzcik, 1980: Figure 8)

and the unit-labor requirement is the highest of the three types of technology. The annualized capital charge is assumed to be negligible in proportion to the annual labor cost, so the average cost curve is nearly constant over most of the volume range. The curve labeled "mass production" represents a capital-intensive, highly automated system with relatively small unit-labor requirements, typical of that used to mass produce a standard product design. The annualized capital equipment charge is the highest among the three technologies. Average cost is very high at small volumes because of the large fixed-capital requirements, but it decreases sharply (by a factor of 1/output) as the volume of output increases. The curve labeled "batch production" represents a semiautomated production technique which is typical of that used when products are manufactured in medium-sized batches. In comparison with the custom- and mass-production technologies, it represents an intermediate case where average cost is not completely dominated by either variable-labor cost or fixed capital cost. Figure 3-3 shows that custom-, batch-, and mass-production methods are the cost-minimizing choices in the low-, medium- and high-volume ranges, respectively. Each of the three technologies is the cost-minimizing choice only within the volume range for which it is intended, but is an inefficient choice outside of its intended range. In summary, for a given product manufactured in a wide range of volumes, unit cost decreases in a regular fashion as output increases only if one considers the lower envelope of the long-run cost curve, where the optimal, cost-minimizing technology is used for each output level.

3.3. Empirical Study of Unit Cost Versus the Volume of Output for a Typical Metalworking Product

The following is a review of the results of one of the few detailed empirical studies of producing a typical metal product. The purpose of this review is to compare the unit cost of producing the product in small, medium, and large volumes and to examine some of the characteristics of the cost-minimizing technique of production for a given product volume. Lamyai, Rhee, and Westphal (1978), of the Economics of Industry Division of the World Bank exhaustively analyzed the cost of producing one metal component, the headlug for a bicycle (that part of the frame into which the fork and handle bar are inserted). The analysis is based on an extensive study of the man-machine input requirements to produce the headlug under alternative technologies (Nam, Rhee, and Westphal, 1973). Lamyai, Rhee, and Westphal analyze how, under static conditions, the optimal choice of technology is affected by volume, factor prices, and the mode of organization. In their analysis, a system of production is specified by the type and form of the raw material used to produce a particular product. A series of process stages, such as shearing, conveying, blanking, forming, and welding, is associated with each system. Each process stage within the system can be carried out in one of several ways. Each alternative is referred to as a technique. Each technique requires a set of specific capital inputs, referred to as process elements and labor inputs. Lamyai, Rhee, and Westphal (1978: 7) explain that "choice of technology to produce headlugs consists of the selection of a system of production and, within it, of a technique at each process." Although their results are based on the analysis of producing a particular design of one bicycle component, they claim, "There is good reason to believe, however, the results obtained for this product characterize most, if not all, mechanical engineering production" (p. 1).

In their analysis, the mode of organization indicates the degree to which equipment and labor are shared among various process stages in the overall manufacturing process. They analyze two extreme modes of organizing production, designated as full-sharing and no-sharing. Under full-sharing, process elements and labor are assumed to be completely divisible; that is, units of each can be hired individually for fractional units of time at the corresponding fraction of the charge per unit of time. Under no-sharing, process elements are recognized as being indivisible, with the further assumption that there is absolutely no sharing of them among process stages; that is, process elements and labor are charged for the entire period of processing time even though they are used in a particular task for only a fraction of that period. The mode of organization is a critical parameter in the Lamyai, Rhee, and Westphal (1978, 1982) studies, since

estimates made under full and no sharing give lower and upper bounds respectively for the minimal cost of production given that the choice of technology is optimized with respect to the organization of production. That is, production cannot be organized such that minimal production cost is less than that under full sharing or more than that under no sharing. (1982, p. 14)

The relevance of their work to the study at hand is their examination of how minimal production costs and characteristics of the chosen technology vary across different output levels for a *given* mode of organizing production.

Lamyai, Rhee, and Westphal's choice of the cost minimizing technology is determined by considering *all* inputs in the production process. Direct labor cost, capital equipment cost, and special tooling cost are calculated under appropriate cost-accounting assumptions for the full-sharing and no-sharing modes of organization. Capital cost includes annual interest and depreciation charges on machines and equipment, as well as special fixture cost. Other direct and indirect costs in their analysis include interest charges on inventory, raw material cost, utility cost, maintenance cost, building and structure cost, and overhead cost for day-to-day management. There is no difference in the computation of these costs between the full- and no-sharing cases. Value added, defined as the difference between the value of output and the cost of purchased material inputs, is used as the measure of production cost.[5]

Part of the Lamyai, Rhee, and Westphal (1978) results are shown in Tables 3-1 and 3-2. These tables present full-sharing and no-sharing modes of organization

5. Value added includes all employee compensation, indirect business taxes, and property-type income. Property-type income includes all forms of profit, net interest payments, capital-consumption allowances, transfer payments, and subsidies. See Kendrick, 1972.

Table 3-1: Effects of Scale and Wage Rates on Production Cost for Bicycle Headlug: Full-Sharing Case

Dependent Variable	Regressors			
	Constant	Wage[a]	Scale[a]	R^2
Value added per unit of output	5.928	0.446 (16.52)	−0.362 (−19.05)	0.909
Direct labor per unit of output	6.132	−0.335 (−5.32)	−0.586 (−13.02)	0.766
Capital services per unit of output	4.157	0.223 (5.07)	−0.187 (−6.03)	0.498
Capital stock per unit of output	12.85	0.184 (5.58)	−0.738 (32.09)	0.945
Capital services per unit of direct labor cost	−1.97	0.559 (5.48)	0.400 (5.56)	0.498
Process-element utilization	−8.680	0.032 (1.39)	0.548 (34.25)	0.949

Source: Lamyai, Rhee, and Westphal, 1978, p. 44, Table 3.
Number of observations = 64.

[a]$(\ ) = \text{t-ratio} = \dfrac{\text{estimate}}{\text{standard deviation of estimate}}$

THE REGRESSION EQUATION IS
$Ln(Y) = b_0 + b_1 \text{ wage} + b_2 \text{ scale}$

WHERE
Y = dependent variable
wage = ratio of average hourly wage to interest rate, relative to that ratio observed in market prices
scale = number of headlugs produced per year

separately. Each table shows a regression equation relating the minimum value added per unit, or a component of value added to a given wage index, and scale of production. The lower (cost-minimizing) envelope of the long run cost curve is approximated by the following regression equation:

$$Ln(Y) = b_0 + b_1 \text{ wage index} + b_2 \text{ output} \qquad (3.1)$$

where

Y = dependent variable
wage index = ratio of average hourly wage to interest rate, relative to that ratio observed in market prices.[6]
output = number of headlugs produced per year
       = 250, 500, 1000, 5000, 10,000, 25,000, 50,000, 100,000, 250,000
$b_1 = d\, Ln(Y) / d\, Ln(\text{wage index})$
$b_2 = d\, Ln(Y) / d\, Ln(\text{output})$

Other forms of the cost function were tested, but the most convenient and simple form, shown here, gives the best results. Since natural logarithms are used, the slope parameter, b_1, measures the percentage of change in unit cost for

6. In other words, a factor price ratio of two means that the ratio of direct hourly wages to interest rates would be twice the ratio that was observed in the market when the data set was compiled.

Table 3-2: Effects of Scale and Wage Rates on Production Cost for Bicycle Headlug: No-Sharing Case

Dependent Variable	Regressors			
	Constant	Wage[a]	Scale[a]	R^2
Value added per	9.019	0.427	-0.524	0.939
unit of output		(15.25)	(-26.20)	
Direct labor per	5.445	-0.292	-0.339	0.616
unit of output		(-5.12)	(-8.47)	
Capital services per	3.960	0.167	-0.249	0.475
unit of output		(3.21)	(-6.73)	
Capital stock per	9.272	0.313	-0.591	0.868
unit of output		(6.95)	(-18.49)	
Capital stock per	3.828	0.605	-0.252	0.467
unit of direct labor cost		(6.30)	(1.32)	
Process-element	-5.390	-0.146	0.345	0.854
utilization		(-5.41)	(18.18)	

Source: Lamyai, Rhee, and Westphal, 1978, p. 44, Table 3.
Number of observations = 64.

$$^{a}(\ ) = \text{t-ratio} = \frac{\text{estimate}}{\text{standard deviation of estimate}}$$

THE REGRESSION EQUATION IS

$Ln(Y) = b_0 + b_1 \text{wage} + b_2 \text{scale}$

WHERE
Y = dependent variable
wage = ratio of average hourly wage to interest rate,
           relative to that ratio observed in market prices
scale = number of headlugs produced per year

a 1 percent increase in the factor price index, and the parameter, b_2, measures the percentage of change in unit cost for a 1 percent increase in the quantity of output. The parameter b_2 is referred to as the scale elasticity, since it measures the elasticity of output with respect to the scale of production. The intercept parameter, b_0, has no particular economic interpretation.

In Tables 3-1 and 3-2 the summary of regression results with respect to scale indicates that for both modes of organization

—the elasticity of value added per unit of output is negative.

—the elasticity of direct labor cost per unit of output is negative.

—the elasticity of capital cost per unit of output is negative.

—the elasticity of the degree of utilization of machines and equipment is positive.

These empirical results are all consistent with the previous discussion on the shape of the long-run average cost curve.

The purpose of reviewing these results in this chapter is to supplement the discussion of the characteristics of low-, medium-, and high-volume production. In accordance with this purpose, the differences between those elasticity estimates derived under the assumptions of full-sharing and those under no-sharing are not discussed further, with the exception of the estimates of the ratio of capital to labor, which are discussed later. The point of interest here is that, under both modes of organization, the optimal values of value added per unit of output, direct labor cost per unit of output, and capital cost per unit of output decrease as the scale of production increases, and the rate of process-element utilization increases as the scale of production increases. However, the estimates vary for the absolute magnitudes of the corresponding elasticity, depending on the mode of organization.[7]

As the scale of production increases, the ratio of value added to unit of output decreases, because its largest components—the ratios of direct labor cost to unit of output and of capital cost to unit of output—both decrease. The relationship between capital cost per unit of output and the scale of production is discussed first. With respect to output in both the full-sharing and no-sharing cases, the estimated elasticity of the ratio of capital cost to unit of output decreases, assuming the appropriate definition of capital cost is used.[8] Assuming the cost-minimizing system of technology is used for a given level of output and factor price

7. For a given volume and factor price ratio, value added per unit of output under the assumptions of full-sharing is always less than under the assumptions of no-sharing. The primary reason for this difference, briefly, can be related to the fact that capital is completely divisible in one case and indivisible in the other. In the full-sharing case, the requirements for a given machine are determined from the total requirements for that type of process *across* separate tasks, or across product modules, whereas in the no-sharing case, the requirements for a given machine are determined from the one specific task to which that machine is dedicated. Because of the pooling of requirements across tasks, output requirements for machines in the full-sharing case are greater than in the no-sharing case for any given level of output. Because of the greater processing requirements, more specialized machines with larger capacities are used in the full-sharing case. The significance of this difference stems from the fact that the increase in the cost of a machine is not proportional to an increase in its capacity. That is, machines with larger capacities do not cost proportionately more, and the amortized machine cost per unit of output is lower than for machines with smaller capacities. In addition, machines in the full-sharing case are always fully utilized (by definition), spreading the cost of the machine over the maximum number of units. In the no-sharing case, machines are not always fully utilized. Less than full utilization increases capital cost, since the yearly amortized cost is divided by the level of machine utilization to fully recover the cost of the machine per period. One of the main conclusions of (Lamyai, Rhee, and Westphal (1978: 57) is that their results "suggest that production organization is as important as, if not more important than, economies of scale given the form of organization."

8. The appropriate method of calculating capital cost differs under the conditions of full-sharing and no-sharing because of the differences in the underlying assumptions associated with the two conditions. In the full-sharing case, machines can be rented for fractional units of time at the corresponding fraction of charge per unit of time, so the process element charges per unit of time employed are independent of the volume of processing. Since capital required for a given task can be momentarily rented, the measure of capital charges that is consistent with the assumptions of full-sharing is rentals on capital used, or the value of capital services. In the no-sharing case, a machine's time cannot be divided into fractional units, so capital costs are charged for the entire period of process-

index, a percentage of increase in the level of output results in a percentage of decrease in the ratio of capital cost to unit of output.[9] Capital cost per unit of output decreases as the level of output increases for two reasons. First, whatever type of machine is used, the level of its utilization—defined by Lamyai, Rhee, and Westphal (1978) of time as the amount that it is actually employed divided by the amount of time that it is available, increases as the level of output increases. Second, at the point when production levels require the purchase of larger-capacity machines, their cost increases by an amount proportionately less than their increase in capacity. That is, machines of larger capacities do not cost proportionately more than machines of smaller capacities.

The estimated elasticity of direct labor cost per unit of output is negative for both the full-sharing and no-sharing cases. As the scale of output increases, the direct labor cost required to process a unit of output decreases. The primary reason for this is that the cycle time per piece decreases as the number of pieces produced increases. This decrease happens for two reasons. First, for a given type of machine, the time required to process a piece, as well as to load and unload the piece, depends on the amount of special tooling (such as jigs and fixtures) used, and on how the controls on the machine are adjusted. The cycle time per piece can be shortened by taking more time to set up the machine and the accessory tooling. However, it is not worth taking the extra time to set up the machine to shorten the cycle times if only a few parts are being made. (It would be faster to make the parts with slower cycle times than to perform a lengthy set-up.) Second, even for a given machine, the labor time required per part decreases as the number of parts processed increases. As volume requirements continue to increase, machines with higher operating rates are substituted for machines with lower processing rates, further reducing labor time per part as the volume of production increases.

Both unit capital and unit labor inputs decrease as the scale of production increases. The relative rate at which unit capital and unit labor requirements

---

ing, even if the machine is used for only a fraction of that time period. The measure of capital charges that is consistent with the assumptions of no-sharing is the annualized investment cost of the machines plus a return on investment, or the value of the capital stock. Capital required for a given task must be bought, even if the same machine is already owned and is being used for another task. For the case of full-sharing, the capital cost component of value added is measured in terms of rentals for capital services. For the no-sharing case, the capital cost component of value added is measured in terms of the annualized cost of buying the capital stock.

9. Alternatively, one could examine the relationship between capital costs and output in total terms, as opposed to unit terms. The estimates of the elasticities of capital cost in total terms are obtained by adding unity to the estimates of the elasticities of capital cost in unit terms with respect to output. Total capital requirements increase as the volume of output increases, but a 1 percent increase in output requires less than a 1 percent increase in capital cost. This is also the case for total requirements of value added and for direct labor.

decrease determines what happens to the ratio of capital to labor as the scale of production increases. It is typically assumed that the ratio of capital to labor increases as the scale of production increases, because as volume requirements increase more automated equipment is substituted for production labor. However, the regression results of the Lamyai, Rhee, and Westphal (1978) analysis show that

—the elasticity of capital cost to direct labor cost with respect to scale is positive for the case of full-sharing.

—the elasticity of capital cost to direct labor cost with respect to scale is negative for the case of no-sharing.

In the full-sharing case, the ratio of capital to labor increases, primarily because of the reason given above—automated equipment is substituted for production labor. Why does the ratio decrease in the case of no-sharing? As volume requirements increase, they can be met by utilizing the existing machine more fully and by adding as many additional labor hours as are required. Thus, higher levels of output can be produced by using the same input of capital but utilizing it more fully with additional labor. This decrease in the ratio of capital to labor occurs in the no-sharing case, because capital is assumed to be completely indivisible and must be rented for an entire period, even if it is required for only a fraction of that period. The important conclusion about the difference between the ratio of capital to labor in the full-sharing case and in the no-sharing case is that the behavior of this cost-minimizing ratio is complex. It is more difficult to generalize about the relationships between the ratio of capital to labor and the scale of production than to generalize about the relationship between unit cost (or unit capital cost or unit labor cost) and the scale of production.

For both modes of organization considered in the Lamyai, Rhee, and Westphal study, the elasticity of unit cost with respect to scale is negative, indicating that unit cost decreases as the scale of production increases. In the full-sharing case, the estimate of the elasticity of unit processing cost with respect to scale is -0.362, with a 95 percent probability that the true value of the scale parameter lies within the interval $< -0.400, -0.324 >$. In the no-sharing case, the estimate of the scale elasticity is -0.524, with a 95 percent probability that the true value of the scale parameter lies within the interval $< -0.564, -0.484 >$. Both estimates have relatively small standard deviations and are highly significant in a statistical sense. These empirically derived average cost curves for the headlug, based on detailed engineering analyses, have the same basic shape as the two hypothetical cost curves discussed earlier in Figures 3-1 and 3-3. Both theoretical and empirical studies concur on the shape of the long-range cost curve. The processing cost for a particular product decreases as volume of production increases, assuming that cost-minimizing technology is used at each level of output.

3.4. Differences in Unit Cost between Custom, Batch, and Mass Production

The approximate magnitude of the difference between the unit cost of a product that is custom, batch, or mass produced is examined here. The scale elasticity of -0.362 from the Lamyai, Rhee, and Westphal (1978) full-sharing case implies a 12-fold difference between the unit cost (value added per unit) of small-batch production (250 units) and very large batch production (250,000 units). The scale elasticity of -0.524 for their no-sharing case implies nearly a 40-fold difference between the unit cost of small-batch production and very large batch production.

Another estimate of the difference between the unit cost of low- and high-volume production can be inferred from Cook's (1975) illustrative cost curve in Figure 3-1. Since the relative cost per part declines nearly linearly as volume increases, and since the graph is plotted on a log-log scale, the relative unit cost can be approximated by

$$Ln(U) = b_0 + b_1 Ln(O) \tag{3.2}$$

where

$U = P/O$ = average cost per unit
P = processing cost = (production labor + equipment + tooling) cost
O = output = number of units produced
$b_1 = d\,Ln(U)\,/\,d\,Ln(O) \tag{3.3}$

Processing cost, as defined by Cook, includes most but not all of the components of value added. Only the production labor component of total labor cost is counted. Capital costs are limited to tooling cost, machine depreciation, and annualized charges to repay the investment in equipment. Thus, this measure of processing cost is not identical to the value-added measure used by Lamyai, Rhee, and Westphal.

Two estimates of the slope and intercept parameters for the straight line in Equation 3.2, based on two different pairs of points in Figure 3-1, are shown below:

	Output	Unit Cost	
	1	100	
Lower bound	1×10^6	1	(3.4)
	$Ln(U) = 4.604 - 0.333\,Ln(O)$		
	1	500	
Upper bound	1×10^6	1	(3.5)
	$Ln(U) = 6.215 - 0.450 Ln(O)$		

The lower bound equation assumes both a log-linear relationship between unit cost and output and that the line passes through the points $(100, 1)$ and $(1, 1 \times 10^6)$. Under these assumptions, Cook's claim that the unit cost in one-of-a-kind production exceeds unit cost in the most efficient mass production by a factor of 100 implies a scale elasticity of -0.333. A factor of 500 in the difference between unit cost in one-of-a-kind production and unit cost in mass production implies a scale elasticity of -0.450. For comparison purposes, in the Lamyai, Rhee, and Westphal study, the expected value of the scale elasticity for the headlug under full-sharing is -0.362, with a 95 percent probability that the true value of the parameter lies within the interval < -0.400, -0.324 >. The lower bound estimate derived from Cook's curve for a typical machined product falls within the estimated range of the scale elasticity for the headlug when produced under the full-sharing mode of organization. The expected value of the scale elasticity for the headlug under no-sharing is -0.524, with a 95 percent probability that the true value of the parameter lies within the interval < -05.64, -04.84 >. When the headlug is produced under the no-sharing mode, the range for the scale elasticity lies above the upper bound estimate implied from Cook's curve. Full-sharing and no-sharing are both abstractions intended to characterize production cost under two extreme modes of organizing equipment and labor. Seldom are factories organized in a way that corresponds exactly to either of these extremes. Factories may come close to one extreme or another, but more often than not fall somewhere in the middle. Consider the interval < -0.564, -0.324 >, formed from the lowest point of the confidence interval for the no-sharing case and the uppermost point of the confidence interval for the full-sharing case. It is interesting that this interval, which defines the range of the scale elasticities under the two extreme modes of organization, brackets the high and low estimates of the scale elasticity derived from Cook's curve, which is based on general observations of how technologies are actually organized.

Based on Equations 3.4 and 3.5, which approximate lower and upper bounds for Cook's cost curve, the difference between unit cost for producing 250 units and 250,000 units per year ranges from a factor of 10–22. Cook's earlier assertion—that for a typical machined product the unit processing cost for the most efficient mass production operations is 10–30 times lower than the unit cost for medium-volume batch operations—is generally consistent with the empirical results reported by Lamyai, Rhee, and Westphal. If Cook's stylized cost curve is valid, then one should view as a reasonable estimate his assertion that at 1 million units per year unit processing cost using the most efficient mass-production techniques is 100–500 times lower than if the product were produced as one-of-a-kind.

3.5. Methodology for the Identification of Custom-, Batch-, and Mass-Production Industries

Most of the value added in the metalworking sector is accounted for by products which are batch produced is a claim that has been in the literature many times (Anderson, 1972; Cook, 1975; Ardnt, 1977; and Thomson, 1980). It is worthwhile to examine the validity of this claim, since it is often used to support the assertion that there is a widespread and urgent need for the use of robotics and of other types of "flexible" automation.

If each metalworking industry could be identified as being composed of predominantly custom, batch, or mass producers, it would be straightforward to calculate the proportional distribution of value added among the modes of production. However, there is no known procedure for classifying SIC manufacturing industries in this way. In the absence of a useful classification procedure, previous estimates of the distribution of value added among the modes of production have been based on the subjective judgement of experienced manufacturing analysts. In order to obtain a more reliable estimate of this distribution, it is necessary to estimate which type of producer—custom, batch, or mass—compose each of the metalworking industries.

The ideal way to estimate how value added and the total value of output are proportioned among the modes of production would be to survey all establishments within the metalworking industries. Each establishment could be classified as a custom-, batch-, or mass-production facility, based on the average batch size for each product and on the type of equipment used. Given the value added for each establishment, the distribution of it and the total value of output among the modes of production could be calculated in a straightforward fashion. However, the large number of metalworking establishments makes a classification procedure based on a comprehensive survey very expensive and difficult to carry out. There are over 107,000 establishments which compose the industries within SIC 34–37. Of these establishments, over 45,000 have more than 20 employees. The Bureau of the Census surveys most of these establishments to compile the various types of data it publishes for manufacturing industries in the *Census of Manufactures*. In addition to publishing input and output statistics, the census also provides a written description of each industry. Unfortunately, with only several exceptions, the census does not attempt to classify industries according to whether they are composed mostly of custom, batch, or mass producers. In some instances, the industry descriptions mention if the products made within the industry are usually manufactured on a job or order basis. These descriptions identify only several of the industries which are composed of custom and small-batch producers and are therefore too limited to be useful in classifying all industries.

Suppose a measure of production cost and of output were available for each metalworking industry within SIC 34–37. Average cost within each industry would be defined by the ratio of total production cost to the units of output produced. Since there are important differences in the composition, size, and complexity of each industry's primary and secondary products, the measure of output for each industry would have to be converted into standardized units so that measures of output could be compared meaningfully across the various industries. Given the almost infinite variety of products manufactured within the metalworking industries, differing in size, shape, material composition, tolerances, and performance-related characteristics, it is not obvious how a standardized unit of output could be defined and constructed to allow meaningful cross-sectional comparisons. One would have to assume that a standardized unit of output could be defined and measured, ignoring for the moment all of the difficulties involved in constructing such a measure. Assume further that each producer within each industry is indeed using the cost-minimizing mode of production for its level of output, so that differences in costs are not due to mismatched comparisons between cost-minimizing technologies and technologies that might actually be in use. Also assume that, within each industry, the value of output *is* not evenly distributed among custom, batch, and mass producers, implying that there is a dominant mode of production used for most of the output within the industry.[10] If the correspondence between the mode of production, the level of output, and the level of unit cost that is maintained when a particular product is produced at different levels of output were still to hold under these assumptions, the case would be:

—Industries with the highest levels of unit cost and the lowest levels of output are composed mostly of custom and small-batch producers,

—Industries with the lowest levels of unit cost and the highest levels of output are composed mostly of mass producers,

—Industries with medium-range levels of unit cost and medium-range levels of output composed mostly of batch producers.

A measure of average cost per unit for each industry requires data on the total cost of production and on the number of units produced. Several measures of production cost are readily available for each industry. The *Census of Manufactures* provides data on production labor cost, total labor cost, value added, and value of shipments. The value of total output can also be closely approximated by adjusting the value of shipments by the change in inventory. Value added is used as the measure of production cost. It includes both labor and capital cost, and is therefore more comprehensive than just labor cost. Value added does not include the cost of purchased materials, which is a desirable exclusion here because material costs are used later in the calculation of an output measure. Data on the units of output produced are provided for only some product classes within some industries. Even if data on the number of units produced were avail-

10. If the unit of observation were an individual establishment, as opposed to an industry composed of many establishments, this last assumption would not be necessary. However, the subsequent analysis is all based on data from industries.

able for all industries, a comparison of cost per unit of output across industries would not be meaningful because of the large differences in the attributes of the products manufactured by each industry. Without a method for converting each industry's output to a standardized unit, a comparison of output and of unit cost cannot be made across industries, eliminating the possibility of using these measures as a means of inferring the dominant mode of production. In most economic applications, dollars are used as the standard unit of measure, and comparisons among industries are made by measuring all inputs and outputs in dollar values. This strategy is not applicable here, since a surrogate measure of the scale of production is required. If the dollar value of output were used, there would be problems in distinguishing between industries making smaller quantities of higher-priced products and industries making larger quantities of lower-priced products. A physical measure of output is required.

One attribute common to all products is mass. In principal, a given volume of output can be converted into an equivalent number of pounds. Measuring output in pounds simplifies the problem of aggregating the heterogeneous mix of products made within each industry, and pound measures for each industry are easily compared across industries. Using pounds of output as the standard unit of output is both conceptually and operationally simple, and as such has an intuitive appeal. Given pounds as the standard unit of output, average cost within each industry is obtained by dividing value added by pounds of output. These measures of average cost and of output are used to provide a basis for a procedure to classify industries according to the dominant mode of production. Industries with the highest levels of production cost per pound of output and with the fewest pounds of output are classified as being composed mostly of custom small-batch producers. Industries with the lowest levels of production cost per pound of output and with the most pounds of output are classified as being composed mostly of mass producers. The remaining industries, those with medium-range levels of production cost per pound of output and with medium-range levels of pounds of output, are classified as being composed mostly of batch producers. The assumption behind these general criteria for classifying industries is that there is an "ideal" correspondence between mode of production, cost, and level of output which underlies the shape of the long-run cost curve for a particular product (in a particular factory). The key assumption made here is that these correspondences are maintained even in a comparison across industries that relies on an unconventional measure of output.

When average cost and output are compared *across* different industries, there are several important factors which might distort the ideal correspondences between the level of output, the level of unit cost, and the mode of production used. If the distortions were large enough, the measures of average cost per pound and of pounds of output might not properly reflect the dominate mode of production within an industry. The principle problem is that when more than one industry is included in the analysis, it is not possible to attribute to scale effects all the differences in average cost per unit across those industries.

The analysis of the average cost curve for a particular product, discussed earlier in this chapter, represents a situation where the attributes of the product itself (size, shape, material composition, tolerances, performance-related characteristics) are assumed to remain constant as the level of output increases. Since the product does not vary, the differences in average cost per unit at each level of increased output are due to various types of economies-of-scale effects. In the cross-sectional comparison presented here, measures of average cost and of output are computed for each separate industry in the analysis. Since products vary in complexity across industries, differences in the nature of the processing requirements associated with this variety of product complexity will affect the cost required to process a given amount of output. This introduces a second major reason for the variation of average cost per unit across industries. Since requirements for labor and/or capital inputs typically increase as the complexity of the product increases, it need not be the case that an industry with a unit cost higher than another produces fewer pounds of output. Differences in the nature of the processing requirements between the two industries could cause such an irregularity.

Given the labor requirement per unit of output and the level of output, differences in wage rates across industries would also affect the magnitude of production cost, and hence the level of unit cost. Production labor cost is only a proportion of total production cost, and for the most part, the variation in wage rates across industries is not extremely wide. Taking these facts into consideration, the extent to which differences in wage rates account for interindustry differences in average unit cost is probably small in comparison with the effects of interindustry differences in the nature of processing activities.

When industries are used as the unit of analysis, differences in the number of establishments within each industry could also introduce noise into a comparison of average cost versus output. Based on their primary product, individual manufacturing establishments are grouped together to define an industry. An industry's output is the aggregated total of the output of these individual establishments; it depends on the output per establishment, and on the total number of establishments within the industry. An industry with a small number of establishments, each producing a large volume of output, could have the same or an even lower total output than an industry with a large number of establishments, each producing a small volume of output. If this were the case, the first industry (few establishments, large output per establishment) would have a lower unit cost than the second one (many establishments, small output per establishment), even though the total output of the first industry is less than that of the second. Defining the measure of output as pounds of metal per number of establishments rather than simply the total number of pounds of metal processed will eliminate this problem.

Pounds of output produced by each industry are not directly measured; they must be estimated. Because of the limitations of available industrywide data, pounds of output must be approximated from the available information on material *inputs* purchased by each industry. This presents a problem because metal removal is one of the primary means of production in the metalworking industries. If material is removed from metal bars, plates, and sheets to shape an industry's product, then the weight of the output will be less than the weight of the purchased material inputs. In fact, the difference between the weight of the purchased materials and the weight of the final product is a good indicator of the complexity of the production process. Given available data sources, there is no way to obtain a measure of the weight of the final output, so one has to settle for a measure of the pounds of metal inputs. Even with this compromise, there are still problems with obtaining a reliable measure of the pounds of material inputs used. First, while the *Census of Manufactures* lists the total dollar value of materials purchased by each industry, only part of the total is "specified in kind," showing a subtotal for each specific type of material purchased.[11] Thus, not all of the materials can be used in the estimate of pounds of inputs purchased. Second, the percentage of total materials purchased that are specified in kind varies across industries, so the material coverage for each industry is not the same. Third, and most important, for some categories of specified materials, the amount purchased is given only in dollar values, making it necessary to estimate the number of pounds of these materials received. For the other categories of materials specified in kind, the number of pounds of material received by all establishments within the industry, along with the dollar value of the purchase, is given by the *Census of Manufactures*. The proportion of materials for which the pounds of inputs must be estimated and the procedure for making this estimate are explained below.

A breakdown of the materials purchased by the industries in major groups SIC 34–37 is shown in Table 3-3. Purchased metals are subdivided into two categories, basic metal inputs and processed metal inputs. The term *basic metals* refers to inputs of "raw" metal stock—steel, brass, and aluminum, in the form of bars, billets, sheets, strips, plates, pipe, tubes, etc., as well as castings and forgings made of the three basic metals. The term *processed metals* refers to inputs which are themselves the products of metalworking industries. In gen-

11. Throughout this study, the term *material purchases*, or *material cost*, refers to the direct charges paid for all raw materials, as well as for semifinished goods, parts, components, containers, scrap, and supplies put into production or used for operation during the year, including freight charges and other direct charges incurred by the establishment in acquiring these materials. The following items are *not* included when material cost is discussed: (1) electric energy purchased, (2) fuels consumed for heat, power, or electricity, (3) materials bought and subcontracted out to be worked on by other manufacturing establishments (contract work), and (4) materials bought and resold in the same condition (resales). For the most part, the raw materials and semifinished goods included here are physically embodied in the product itself.

Table 3-3: Breakdown of Material Inputs for Major Groups SIC 34–37, 1977

Total Material Cost[a] (in millions of dollars)	34	35	36	37
	39,703.9	49,809.3	34,739.5	99,368.3
Material Inputs	Percentage of Total Material Cost			
Specified Materials				
Metals, total	65.2	57.8	37.1	74.6
Basic metals	61.5	27.5	13.9	14.8
Processed metals	3.7	30.3	23.2	59.8
Other[b]	2.4	2.2	4.8	3.3
Unspecified materials	32.4	39.9	58.1	22.1
Total material cost	100.0	100.0[c]	100.0	100.0
Ratio of Processed Metal Cost to Basic Metal Cost	0.06	1.10	1.67	4.04

Source: Compiled from the 1977 Census of Manufactures (Bureau of the Census, 1981c,d).

[a]Includes the direct charges paid for all raw materials, as well as for semifinished goods, parts, components, containers, scrap, and supplies. The following items are not included: (1) electric energy purchased, (2) fuels consumed for heat, power, or electricity, (3) materials bought and subcontracted out to be worked on by other manufacturing establishments (contract work), and (4) materials bought and resold in the same condition (resales).

[b]Other specified materials include paints and varnishes, plastic resins, plastic products, fabricated rubber products, rubber belts and hoses, tires and inner tubes, glass and glass products, and boxes and containers.

[c]Does not add to this total because of rounding.

eral, these products are basic metals which have been further processed within the metalworking industry. For example, the turbine industry (SIC 3511) purchases bolts, which are the primary product of the bolt, nut, rivet, and washer industry (SIC 3452). The delivered cost of the purchased bolt is part of the cost of processed metal inputs. In general, any purchased input which is the primary product of an industry within the major groups SIC 34–38, including the industry of the purchaser, is counted as a "processed metal" input.

There are large differences in the ratio of basic metal cost to total material cost across the four major groups of metalworking industries. See Table 3-3 for specific information. Most of the difference between total material cost and total metal cost is accounted for in the table by the category labeled "unspecified materials." The Bureau of the Census materials survey requests that the responding establishments itemize material inputs in only a limited number of prespecified categories. In many industries these prespecified categories account for most of the total material inputs, and unspecified materials account for a small proportion of total material cost. However, for some industries, especially those in SIC 36, the prespecified categories account for only a small fraction of the total material input, leaving most of the material unidentified. For a small number of industries, the Bureau of the Census has conducted a supplemental survey to obtain information on the use of the unspecified materials.[12] The supplemental survey shows that a large fraction of these unspecified materials are processed metal inputs from other metalworking industries.

For most metalworking industries, metal inputs account for most of the value of materials which are specified in kind. Only metal inputs, including both basic and processed metals, are used to estimate the pounds of materials used by each industry. This considerably simplifies the laborious task of estimating the pounds of materials used, and in most cases excludes only minor quantities of materials. If purchases of basic metals and processed metals account for less than 50 percent of an industry's total material purchases in dollar terms, the industry is excluded from the analysis. This rule excludes 6 industries from SIC 34, 3 industries from SIC 35, 18 industries from SIC 36, and 4 industries from SIC 37. The remaining industries are appropriately designated as metalworking industries, since specified metal inputs account for at least half of total material cost.

For the most part, the number of pounds of material purchased is listed in the *Census of Manufactures* for all categories of the basic metal inputs. However, in some industries, the pounds purchased for a particular basic metal input might not be listed, either to keep from disclosing proprietary data or because the estimate did not meet the publication standards of the census. In the cases where the figure was not disclosed or was suppressed, it was estimated. The estimation procedure is guided by two principles. First, quantities estimated at lower levels of aggregation must be consistent with the information given at higher levels of aggregation. For example, for a given quantity, the sum of the three-digit SIC group subtotals must add to the total for the two-digit SIC major group. Similarly, the sum of the quantity for the four-digit industries must add to the subtotal for the three-digit group. Second, where it was necessary to allocate a given dollar's worth of material inputs across two or more four-digit industries within the same three-digit group, the allocations were based on input coefficients from the 1972 input-output table of the U.S. Department of Commerce (1979).

Material inputs purchased from other metalworking industries (processed metal inputs) are given only in dollar amounts, making it necessary to estimate the pounds of basic metal embodied in a dollar's worth of processed metal input. The procedure is summarized here. The estimate of basic metals purchased by each industry is used to approximate the metal content of processed metal inputs. The total pounds of metal processed by industry i can be estimated by

---

12. Some of the industries included in the *1977 Census of Manufactures* supplemental inquiry (Bureau of Census, 1981b) were radio and T.V. communication equipment (3662), motor vehicles and equipment (all four-digit industries in 371), aircraft equipment (all four-digit industries in 372), ship and boat building (all four-digit industries in 373), and guided missile and space vehicles (3761). For these industries, the supplementary information on purchased material inputs is included in the analysis.

$$m_i = b_i + \sum_j a_{i,j}b_j + \sum_k \sum_j a_{j,k}a_{i,j}b_j + \sum_l \sum_k \sum_j a_{k,l}a_{j,k}a_{i,j}b_j + \dots$$

$$(3.6)$$

where

m_i = pounds of metal processed by industry i

b_i = pounds of basic metal processed by industry i

$a_{i,j}$ = $\dfrac{\$ \text{ sales of industry j to industry i}}{\$ \text{ output of industry j}}$

     = proportion of industry j's output sold to industry i

These equations can be summarized in matrix form:

$$M = [I + A + A^2 + A^3 + \dots]B$$

where the solution to Equation 3.6 is given
by the well known result in input-output theory,

$$M = [I - A]^{-1}B \qquad (3.7)$$

where

M = column matrix of all m_i values. Dimension (n × 1).

I = n × n identity matrix

A = matrix of output coefficients, $a_{i,j}$. Dimension (n × n).

B = column matrix of all b_i values. Dimension (n × 1).

Equation 3.6 is explained with the assistance of Figure 3-4, which illustrates how basic metal inputs are embodied in purchases from other metalworking industries. The pounds of basic metal directly purchased by industry 1 are given by b_1. The proportion of industry 2's output sold to industry 1 is given by $a_{1,2}$. The number of pounds of basic metal purchased by industry 2 is given by b_2, and the other output coefficients, $a_{2,3}$ and $a_{2,4}$, are derived from the sales of industries 3 and 4 to industry 2. Similarly, the proportion of industry 3's output sold to industry 1 is shown ($a_{1,3}$), along with the basic and processed metal inputs used by industry 3. The first summation in Equation 3.6,

$$\sum_j a_{i,j}b_j = a_{1,2}b_2 + a_{1,3}b_3 \qquad (3.8)$$

gives an estimate of the basic metal directly embodied in the processed metal inputs purchased by industry 1. The second summation in Equation 3.6 adds the amount of basic metal inputs embodied in the processed metal inputs of the processed metal inputs. For example, in Figure 3-4, industry 1 buys inputs from industry 2, and industry 2 buys inputs from industry 3. The basic metal embodied in a dollar's worth of purchases from industry 2 depends on b_2, but also on the basic metal embodied in the processed metal inputs purchased by industry 2, which is given by $a_{1,2}a_{2,3}b_3$. In matrix form, Equation 3.7 accounts for all of the indirect contributions of basic metal embodied in the processed metal inputs.

The basic metal directly embodied in the processed metals can easily be calculated from the first summation (Eq. 3.8) without explicitly using the full (150

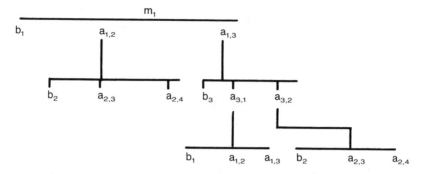

m_1 = total pounds of metal processed by industry 1.
b_1 = pounds of basic metal processed by industry 1.
$a_{i,j} = \dfrac{\$ \text{ of sales from industry j to industry i}}{\$ \text{ output of industry j}}$
= proportion of industry j's output sold to industry i.

Figure 3-4: Calculating the Basic Metal Embodied in Processed Metal Inputs

by 150) matrix of output coefficients, A. Determining all the indirect contributions of basic metals embodied in the processed metal inputs purchased by each industry requires that matrix A be inverted. In this analysis, Equation 3.7 is not solved because of the problems associated with forming and inverting a very large matrix. Only the basic metals directly embodied in the processed metal inputs are included by calculating Equation 3.8 for each industry in the analysis. If industry i's inputs were composed mostly of processed metals, and if the processed metals themselves were composed mostly of processed metals, the estimate of the number of pounds of metal processed would be substantially understated because the indirect chains are not included here.

In summary, an estimate of the number of pounds of basic metal and of processed metal inputs purchased is used as an approximate measure of the pounds of material inputs processed by each industry. The number of pounds of metal processed is also used as a surrogate measure of the pounds of output produced by an industry. Using pounds as the standardized unit of output, cost per unit of output and the units of output are computed for each industry. An industry is classified as being composed mostly of custom, batch, or mass producers according to its position on a graph of average cost versus output. The rules defining this particular classification are based on the theoretical correspondences between the level of average cost, the level of output, and the mode of production. These rules are explained in more detail in Section 3.6. While this scheme is conceptually simple and operationally straightforward, there are several factors which would introduce uncertainty into the analysis, even if a perfect measure of pounds of output were available. When the average cost and the output of different industries are compared, the difference in the levels of outputs is only one of several factors affecting cost variations across industries. In addition to scale effects, differences in product composition and complexity also influence an industry's average cost level. Approximating pounds of output by pounds of material inputs, using only metal material inputs, and estimating the pounds of metal embodied in metal products purchased from other metal-working industries introduces additional sources of noise into the analysis. If the effects of noise are large enough, an industry's position on the graph of average cost versus output will not accurately indicate its true mode of production, and the industry will be classified incorrectly.

3.6. The Relationship between Unit Cost and Output Levels across Metalworking Industries

The underlying determinants of the shape of the average cost curve for a particular product were discussed earlier in this chapter. The Lamyai, Rhee, and Westphal (1978) study shows that, for the case of a particular product, if the cost-minimizing mode of production is used for a given level of output, then

—capital cost per unit of output decreases as the level of output increases.

—direct labor cost per unit of output decreases as the level of output increases.

—value added per unit of output decreases as the level of output increases.

—the level of machine utilization increases as the level of output increases.

The question here is whether or not these ideal relationships between unit cost and the level of output that hold for a particular product are also maintained in the proposed analysis of value added. This question can be tested statistically. To support the use of the proposed output measure, and to support the assumption that an industry's mode of production can be inferred from its position on a graph of value added per pound of metal versus pounds of metal processed, the following should hold true across all industries:

1. Capital cost for equipment and machinery per pound of metal processed decreases as the number of pounds of metal processed increases.

2. Production labor cost per pound of metal processed decreases as the number of pounds of metal processed increases.

3. Value added per pound of metal processed decreases as the number of pounds of metal processed increases.

4. Machine utilization increases as the number of pounds of metal processed increases.

The gross book value of machinery and equipment is used as a surrogate measure of fixed equipment cost. If annual equipment and machinery costs are proportional to the gross book value of equipment and machinery, then this surrogate measure is a reasonable indicator of annual capital equipment charges.[13] Unit labor cost is approximated by the ratio of total production worker wages paid for hours worked to pounds of metal processed. Total production wages do not include supplemental payments for benefit programs. Value added is based on the *Census of Manufactures* estimates and includes all forms of employee compensation, indirect business taxes, and all forms of profit-type income.

Because of constraints on available data, estimates of machine utilization could be constructed only for industries at the three-digit level of aggregation. Machine utilization is defined here as follows:

$$\text{utilization} = \frac{\text{total operator hours available}}{\text{total machine hours available}}$$

where

total operator hours available =

number of operators $\times$ average hours worked per operator per year

average hours worked per operator per year =

total hours worked by production workers/total production workers

total machine hours available =

number of machines $\times$ 8760 hours per year

8760 hours per year =

24 hours per day $\times$ 365 days per year

The numerator of the utilization measure gives the total hours worked by metalcutting and metalforming machine operators during the year. The average number of hours worked per operator per year is derived from Bureau of the Census data for all production workers, including overtime hours but not including hours paid but not worked. The denominator, total machine hours available, is derived from estimates of the number of machines within each industry published in the "12th *American Machinist* Inventory of Metalworking Equipment, 1976–1978" (*American Machinist*, 1978). Theoretically, machines are available 24 hours per day every day of the year, with the exception of a minute fraction of this time used for scheduled maintenance. It is assumed that on average one operator controls one machine tool. With numerically controlled machine tools,

13. An alternative measure of capital cost can be derived by subtracting total labor cost from value added. The nonlabor component of value added includes annual equipment and machinery costs, but also includes interest on buildings and other types of capital, profits, indirect business taxes, transfer payments, and subsidies. Since equipment and machinery expenses compose only a portion of nonlabor value added, the gross book value of machinery and equipment is preferred as a surrogate measure of capital equipment costs, and is used here to test the hypothesis that fixed equipment cost per unit of output decreases as the level of output increases.

one operator might control two or more machines. In such cases, one operator hour would correspond to two (or more) machine hours. This adjustment is ignored because fewer than 3 percent of the machines in the 12th *American Machinist* Inventory are numerically controlled. With some of the larger, special-purpose transfer lines, there is more than one operator per machine. In this case, one operator hour would correspond to less than one machine hour. This adjustment is also ignored because transfer machines constitute less than 1 percent of the machines in the inventory. In total, the exceptional cases where one operator hour corresponds to more or less than one machine hour are few, and probably cancel each other out. Assuming the estimates of the number of operators and the number of machines are correct, utilization, as defined in this manner, provides an upper bound on the average proportion of time that all machines are in use. The ratio measures only the fraction of time operators are available to run the machines, with no consideration of whether the machine is actually being used when the operator is on duty.

There is one serious problem when using a comparison of total operator hours to total available machine hours for estimating levels of machine utilization. This scheme assumes that each machine is equally important in the production process. In other words, each hour of utilized machine time provides the same marginal contribution to output. In general, this is not true. It has been established in several plant-level surveys that a small proportion of the machines, typically the newer, more productive ones, are used to perform the majority of the work.[14] Older machines are typically not discarded, even though they are not used on a regular basis. They are sometimes kept in reserve for surge capacity, or permanently set up to perform unusual types of jobs. While the utilization estimates cited above are correct for the average proportion of time that *all* machines are attended, they probably substantially understate the proportion of time that the primary machines are in operation. Despite these problems, these estimates are still useful for comparing utilization rates across industry groups.

Economics texts typically argue that large-volume producers are more capital intensive than small-volume producers on the grounds that the larger the volume of output, the more opportunities there are to substitute machines for production labor. The discussion earlier in this chapter supports the general assertion that the degree of capital intensity increases as the volume of output increases. However, the study of Lamyai, Rhee, and Westphal (1978) shows that there can be conditions where the cost-minimizing ratio of capital to labor cost does not

14. In a survey of 1,100 grinding machines used throughout many batch-production shops, Carter (1980) found that 17 percent of the machines performed 50 percent of the work and that 47 percent of the machines performed 80 percent of the work. In general, the newer machines, which were more productive, were more fully utilized. In a survey of users of numerically controlled metalcutting machines, Lund (1977) also reports that, in some instances, 85 percent of the work was handled by 15 percent of the machines.

increase as the level of output increases.[15] If the ratio of capital to labor were to increase across industries as the number of pounds of metal processed increases, it would lend additional support to the proposed analysis. However, the reader is cautioned that the ideal relationship between the ratio of capital to labor and the level of output is not as straightforward as the relationship between unit cost (or cost components) and the level of output. Two measures of capital intensity are used to examine changes in the ratio of capital to labor across industries as the number of pounds of metal processed increases. The first measure is the ratio of gross book value of machinery and equipment to annual production labor payments. This measure is used to approximate the ratio of annual capital cost to annual labor cost. The second measure is the ratio of gross book value of machinery and equipment to the total number of employees, which is used to approximate the amount of capital investment per employee.

One would expect that the ratio of salaried labor cost to production labor cost should decrease as the level of output increases. In a factory producing customized products, each product is especially designed to the customer's specifications, whereas all products are identical (or nearly so) in a factory producing standardized products. This suggests that product and tool designers should compose a larger fraction of the work force in low-volume factories than in high-volume factories. Also in factories which produce specialized products in batches, products may take one of many possible routes through the general-purpose machines, whereas in high-volume factories, products follow a predetermined route through special-purpose machines. This suggests that production schedulers, expediters, and production planners should compose a larger proportion of the work force in low-volume factories than in high-volume factories. Together, these two examples suggest that salaried workers (including designers, schedulers, production planners) should compose a larger proportion of the work force in low-volume factories than in high-volume factories. This leads to the conclusion that the ratio of salaried worker cost to production labor cost should decrease as the level of output increases.

These hypotheses are specified in equation form:

Equation Forms	Maintained Hypothesis	
$\ln(k/m) = \beta_0 + \beta_1\ln(m)$	$\beta_1 < 0$	(3.9)
$\ln(l/m) = \beta_0 + \beta_1\ln(m)$	$\beta_1 < 0$	(3.10)
$\ln(va/m) = \beta_0 + \beta_1\ln(m)$	$\beta_1 < 0$	(3.11)
$\ln(k/l) = \beta_0 + \beta_1\ln(m)$	$\beta_1 > 0$	(3.12)
$\ln(k/e) = \beta_0 + \beta_1\ln(m)$	$\beta_1 > 0$	(3.13)
$\ln(u) = \beta_0 + \beta_1\ln(m)$	$\beta_1 > 0$	(3.14)
$\ln(s/l) = \beta_0 + \beta_1\ln(m)$	$\beta_1 < 0$	(3.15)

15. In Lamyai, Rhee, and Westphal's study, the value of the cost-minimizing ratio of capital to labor increased as the level of output increased for the full-sharing case, but decreased as the level of output increased for the case of no-sharing.

where

k/m   = gross value of équipment and machinery/pounds of metal
l/m   = all included production worker costs/pounds of metal processed
va/m = value added/pounds of metal
k/l   = gross value of equipment and machinery/all included production worker costs
k/e   = gross value of equipment and machinery/total employees
u   = utilization = hours worked by machine operators/total available machine hours
s/l   = salaries/hourly production worker wages

The constant and slope parameters for each equation are estimated using the ordinary least squares method for calculating regression coefficients. Since all variables are expressed in natural logarithms, the slope parameter in each equation, β_1, is an elasticity, indicating the percentage of change in the dependent variable for a 1 percent increase in the number of pounds of metal processed. The hypotheses are tested by examining the sign of the output elasticities in the above equations. To support the stated hypotheses, the estimate of the output elasticity must be negative in Equations 3.9, 3.10, 3.11, and 3.13, and positive in Equations 3.12 and 3.14..

Earlier, it was mentioned that, when industries are used as the unit of analysis, differences in the number of establishments per industry can introduce noise into a comparison of average cost (or cost components) versus output. It was pointed out that problems resulting from differences in the number of establishments per industry can be adjusted for by redefining output in terms of pounds of metal processed per establishment. The parameters in Equations 3.9 through 3.15 are first calculated using pounds of metal processed and then pounds of metal processed per number of establishments as the measure of output.

The regression results using pounds of metal processed as the measure of output are summarized in Table 3-4. The results using pounds of metal processed per number of establishments are summarized in Table 3-5. Industries within each of the four major metalworking groups (SIC 34, 35, 36, and 37) are analyzed by pooling the data at the four-digit SIC level of aggregation and at the three-digit SIC level.

3.6.1. TEST OF HYPOTHESES USING POUNDS OF METAL PROCESSED AS THE OUTPUT MEASURE

The results obtained using pounds of metal processed as the measure of output (Table 3-4) are discussed first. For the four-digit data set, the estimate of the output elasticity is negative and highly significant for unit capital cost, unit labor cost, and for value added per unit. In these three cases, there is more than a 99 percent probability that the estimated coefficient is significantly different from zero. These same three output elasticities are also negative and highly significant for the three-digit data set, shown in the lower half of Table 3-4. Based on

Table 3-4: Summary of Regression Results of Cost versus Pounds of Output across Metalworking Industries

Dependent Variable	Constant[a] b_0	Output Elasticity[a] b_1	Significance Level for Output Elasticity	R^2
	Pooled Four-Digit Data Set, SIC 34–37: 101 Four-Digit Industries			
k/m	1.881 (4.36)	−0.423 (9.48)	0.99	47.1
l/m	1.868 (4.82)	−0.446 (−8.15)	0.99	39.6
va/m	3.331 (8.24)	−0.478 (−8.39)	0.99	40.9
k/l	0.109 (−0.63)	0.014 (0.58)	reject	−0.7
k/e	1.545 (6.93)	0.109 (3.46)	0.99	9.9
s/l	0.249 (1.16)	−0.119 (−3.91)	0.99	12.5
	Three-Digit Data Set, SIC 34–37: 27 Three-Digit Industry Groups			
k/m	3.637 (3.75)	−0.539 (−4.84)	0.99	46.3
l/m	3.557 (4.29)	−0.556 (−5.83)	0.99	55.9
va/m	5.661 (6.01)	−0.655 (−6.05)	0.99	57.8
k/l	−0.048 (−0.11)	0.009 (0.18)	reject	−3.9
k/e	1.194 (2.48)	0.142 (2.57)	0.98	17.7
u	−3.270 (−4.70)	0.120 (−1.50)	reject	4.6
s/l	1.898 (3.29)	−0.291 (−4.39)	0.99	41.2

Output = pounds of metal processed.
Reject = reject hypothesis.
[a]( ) = t-ratio for estimate.

the data, the first three hypotheses regarding the relationships between cost and the pounds of metal processed (Eqs. 3.9, 3.10 and 3.11) *cannot* be rejected.

For both the four- and the three-digit data sets, the estimate of the elasticity of capital intensity k/l with respect to output is negligible in magnitude, with a relatively large standard error. This indicates that there is no significant correlation between pounds of metal processed and the ratio of the gross book value of machinery and equipment to production labor cost. However, the estimate of the elasticity of capital intensity k/e with respect to output is positive in magnitude and significantly different from zero, indicating that there is a significant positive correlation between pounds of metal processed and the gross book value of machinery and equipment per number of employees.

The estimate of the elasticity of machine utilization with respect to the level of

Table 3-5: Summary of Regression Results of Cost versus Pounds of Output per Number of Establishments across Metalworking Industries

Dependent Variable	Constant[a] b_0	Output Elasticity[a] b_1	Significance Level for Output Elasticity	R^2
	Pooled Four-Digit Data Set, SIC 34–37: 101 Four-Digit Industries			
k/m	−0.724 (−6.75)	−0.371 (−5.49)	0.99	22.5
l/m	−0.835 (−8.95)	−0.440 (−7.49)	0.99	35.5
va/m	0.401 (3.96)	−0.436 (−6.83)	0.99	31.4
k/l	−0.056 (−1.41)	0.051 (2.06)	0.95	3.1
k/e	2.160 (44.17)	0.156 (5.08)	0.99	19.9
s/l	−0.473 (−9.34)	−0.114 (−3.57)	0.99	10.5
	Three-Digit Data Set, SIC 34–37: 27 Three-Digit Industry Groups			
k/m	−0.634 (3.05)	−0.352 (−2.61)	0.99	18.3
l/m	−0.815 (−4.39)	−0.393 (−3.27)	0.99	27.1
va/m	0.431 (1.89)	−0.390 (−2.65)	0.98	18.8
k/l	0.009 (0.12)	0.021 (0.42)	reject	−3.3
k/e	2.269 (27.14)	0.141 (2.60)	0.98	18.2
u	−2.540 (−27.36)	.279 (4.65)	0.99	44.2
s/l	−0.450 (−3.59)	−0.149 (−1.83)	0.90	8.3

Output = pounds of metal processed per number of establishments.
Reject = reject hypothesis.
[a]( ) = t-ratio for estimate.

output is available for only the three-digit data set. While the estimated elasticity is positive, it is not significantly different from zero. Using pounds of metal processed as the measure of output, the hypothesis that machine utilization increases as the level of output increases must be rejected. The estimate of the elasticity of salaries to production wages s/l with respect to output is also negative and significant, supporting the hypothesis that the ratio of salaried labor cost to production labor cost decreases as the level of output increases.

The one objectionable result in Table 3-4 is that the elasticity of machine utilization with respect to output is not significant (even though it has the proper sign). The rejection of this hypothesis (Eq. 3.14) is important. In the Lamyai, Rhee, and Westphal (1978) study of the bicycle headlug, the degree of machine utilization increased as the level of output increased under both modes of organization studied. Also the results of a study of theoretical capacity in low-, medium-, and high-volume metalworking operations reported by Mayer and Lee (1980a) (Figure 3-5) indicate that utilization levels increase as the level of output increases. Figure 3-5 gives estimates of the proportion of time in a year that a typical plant (low-, medium-, and high-volume) is closed down. These figures imply that high-volume plants shut down nearly 80 days per year because of Sundays, holidays, and planned closings for retooling. Medium-volume plants are closed, on average, 102 days (all weekends), and low-volume plants are closed nearly 125 days out of the year (weekends plus three weeks for holidays and shutdowns). The figures also show that, when open for production, high-volume plants typically operate over 22 hours per day, whereas medium- and low-volume plants are typically scheduled to operate 10.7 hours and 8 hours per day, respectively. The percentage of time in a year that is actually scheduled for production in low-, medium-, and high-volume metalworking plants is deduced from the breakdowns of theoretical capacity in Figure 3-5, and is shown in Table 3-6. The figures in Table 3-6 show the fraction of time scheduled for production to be 22 percent for a low-volume plant on one-shift operation, 32 percent for a medium-volume plant, and almost 73 percent for a high-

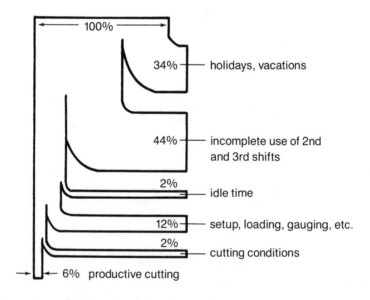

Low-volume manufacturing

Figure 3-5: Breakdown of Theoretical Capacity in Low-, Medium-, and High-Volume Manufacturing (after Mayer and Lee, 1980a, Figure 6)

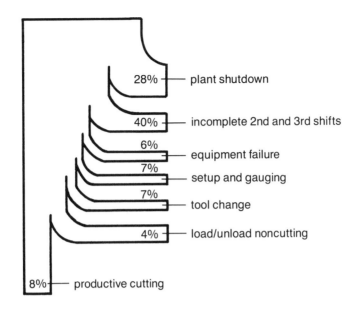

28% ── plant shutdown

40% ── incomplete 2nd and 3rd shifts

6% ── equipment failure

7% ── setup and gauging

7% ── tool change

4% ── load/unload noncutting

8% ── productive cutting

Medium-volume manufacturing

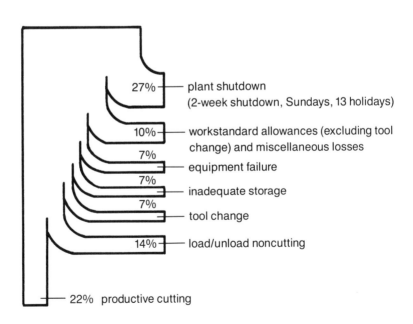

27% ── plant shutdown
(2-week shutdown, Sundays, 13 holidays)

10% ── workstandard allowances (excluding tool
change) and miscellaneous losses

7% ── equipment failure

7% ── inadequate storage

7% ── tool change

14% ── load/unload noncutting

22% productive cutting

High-volume manufacturing

Table 3-6: Estimated Planned Production Time in Low-, Medium-, and High-Volume Metal Fabricating Manufacturing

	High Volume	Medium Volume	Low Volume
Maximum days per year available	365	365	365
Days per year open for operation	286	263	241
Hours per day allotted for:			
Production	22.2	10.7	8.0
Preventive maintenance[a]	1.0	1.0	1.0
Unscheduled (idle) time	0.8	12.3	15.0
Scheduled production time			
proportion of maximum available time	72.6%	32.0%	22.0%

Source: Derived from breakdowns of theoretical capacity in Figure 3-5.
[a]Most factories do not stop production on a daily basis to perform scheduled preventive maintenance. Machines are typically serviced on an "as needed" basis. Assuming one hour per day of scheduled maintenance may even be a high estimate. Major machine overhauls and repairs are typically carried out during scheduled plant shutdowns.

volume plant. Since the total hours worked by machine operators per year are directly proportional to the percentage of time that the plant is scheduled for operation, these figures indicate that the ratio of hours worked by machine operators to total machine hours available should be lowest for custom-production plants and highest for high-volume plants. Machine utilization in Equation 3.14 is defined as the ratio of hours worked by machine operators to the total machine hours available. This being the case, Figure 3-5 and Table 3-6 provide strong evidence for supposing that if output is properly measured the measure of machine utilization used in this study should increase as the volume of output increases. Since the elasticity of machine utilization with respect to the level of output is not significantly different from zero, there is reason to question whether the estimate of pounds of metal processed for each industry is an appropriate surrogate measure of output.

Another result that does not support the proposed measure of output is that the elasticity of the ratio of capital to labor k/l is insignificant. As mentioned earlier, there is more ambiguity in assessing the relationship between this variable and the level of output than with the other variables that are included in the analysis. Also, the elasticity of the other capital intensity measure, k/e, with respect to output *is* positive and significant.

3.6.2. Test of Hypotheses Using Pounds of Metal Processed per Number of Establishments as an Output Measure

When the measure of output is revised to pounds of metal processed per number of establishments (Table 3-5), all of the hypotheses (Eqs. 3.9 through 3.15) are supported by the data. Like the previous results, the estimate of the output elasticity is negative and highly significant for unit capital cost, unit labor cost, and for value added per unit for both the four-digit and three-digit data sets.

It is noted that, while the coefficients of the estimated elasticity for Equations 3.9, 3.10, and 3.11 are all of the proper sign and are highly significant, the standard deviations are somewhat larger than with the previous case.

At the four-digit level, the estimate of the elasticity of capital intensity k/l with respect to output is positive and significant, indicating there is a significant correlation between pounds of metal processed and the ratio of the gross book value of machinery and equipment to production labor cost. This is in contrast to the previous results obtained when pounds of metal were used as the measure of output. However, at the three-digit level, the estimate of the elasticity of k/l with respect to output is still insignificant, although positive. Also, the estimate of the elasticity of capital intensity k/e with respect to output is positive in magnitude and significantly different from zero (with the coefficient having a smaller standard deviation than in the previous case), indicating there is a significant positive correlation between pounds of metal processed and the gross book value of machinery and equipment per number of employees.

The most important difference between the results in Tables 3-4 and 3-5 is that in Table 3.5 the output is divided by the number of establishments. As a result, in that table the estimate of the elasticity of machine utilization with respect to the level of output for the three-digit data set is positive and highly significant. The hypothesis that machine utilization increases as the level of output increases is supported by the data when pounds of metal processed per number of establishments is used as the measure of output. The estimate of the elasticity of s/l with respect to output is also negative and highly significant with the four-digit data set, and is negative and less highly significant with the three-digit data set.

In the results in Table 3-5, all estimated elasticities support the stated hypotheses and are statistically significant. (The one borderline case is the measure of s/l at the three-digit level). This suggests that the preferred measure of output is pounds of metal processed per number of establishments. However, this choice is not made without any reservations. In the regressions of capital cost per unit of output, production labor cost per unit of output, and value added per unit of output against output (Eqs. 3.9, 3.10, and 3.11), the standard deviations of each of the three estimated elasticities are smallest and the goodness of fit measures are best when output is measured by pounds of metal processed. If one were to compare only how unit cost and components of unit cost vary with output, the measure of pounds of metal processed would be preferred over pounds of metal processed per number of establishments. However, in the regressions of capital intensity (measured by either k/l or k/e) and machine utilization against output (Eqs. 3.12 and 3.14), the standard deviations are substantially smaller and the goodness of fit measures are best when output is measured by pounds of metal per number of establishments. If measures of capital utilization and capital intensity are important, then pounds of metal per number of establishments is the preferred measure of output. Since there are strong reasons for believing that as the level of output increases value added per unit of output should decrease *and* the measure of machine utilization should increase, pounds of output per number of establishments is the preferred measure of output.

In summary, a comparison of cost versus the scale of output across a cross section of industries exhibits the same key structural patterns that underlie the shape of the long run unit cost curve for a particular product. Since there is a correspondence between the the level of output, the level of unit cost, and the mode of production used in the long-run unit cost curve for a particular product, it is reasonable to assume that there should also be similar correspondences in a comparison of cost versus output across industries. This is the basic rationale for proposing that the dominant mode of production used within an industry can be inferred from the measures of the level of output and unit cost.

There is still a considerable amount of variation in the unit cost across industries that is not explained by the differences in the levels of output. While the sources of noise do not invalidate the proposed analogy between a theoretical unit cost curve and a comparison of unit cost versus output across industries, a more comprehensive analysis is needed to examine all factors influencing unit cost variations across industries. This is done in Chapter 4.

3.6.3. RULES FOR CLASSIFYING INDUSTRIES BY MODE OF PRODUCTION

Within each major group of industries, the variables cost per pound and pounds of metal processed per number of establishments are partitioned into low, medium, and high levels, as shown in Table 3-7. There are nine (3 × 3) cells if all combinations of levels for both variables are considered. The number of pounds of metal processed per number of establishments increases from cell to cell moving across the table from left to right. Unit cost decreases from cell to cell moving from top to bottom. The top right corner cell for observations with the highest levels of unit cost and with highest levels of output should be empty, as should be the bottom left corner cell for observations with lowest levels of unit cost and lowest levels of output. The empty cells can be explained by considering the idealized long-run cost curve for a particular product, which shows that unit cost decreases as the scale of output increases. Because the highest level of unit cost occurs with the lowest level of output and the lowest level of unit cost occurs with the highest level of output, these two cells should be empty. And they *are* empty, with one exception. There is one industry which has a high level of unit cost and a high level of output. It is believed that the high unit weight of the products of this one industry, engines and turbines, results in a distortion of the output measure. This raises the question of whether differences in unit weight of products across industries exert a larger influence on the measure of output than does the number of units produced. Industries with products of low unit weight having greater measures of output than industries with products of

Table 3-7: Clustering of Industries by Mode of Production

Value Added per Pound of Metal		(m_1) Low (m_2)	(m_3) Medium (m_4)	(m_5) High (m_6)
		Region 1	**Region 2**	
High	(u_1)	CUSTOM PRODUCTION highest levels of unit cost; lowest levels of output	CUSTOM–SMALL-BATCH PRODUCTION highest levels of unit cost; medium levels of output	EMPTY highest levels of unit cost; highest levels of output
	(u_2)			
		Region 3	**Region 4**	**Region 5**
Medium	(u_3)	CUSTOM–SMALL-BATCH PRODUCTION medium levels of unit cost; lowest levels of output	MEDIUM-BATCH PRODUCTION medium levels of unit cost; medium levels of output	LARGE-BATCH PRODUCTION medium levels of unit cost; highest levels of output
	(u_4)			
			Region 6	**Region 7**
Low	(u_5)	EMPTY lowest levels of unit cost; lowest levels of output	LARGE-BATCH PRODUCTION lowest levels of unit cost; medium levels of output	MASS PRODUCTION lowest levels of unit cost; highest levels of output
	(u_6)			

The boundaries for the high-, medium-, and low-cost regions are defined by u_1 through u_6. The boundaries for the low-, medium-, and high-volume regions are defined by m_1 through m_6.

high unit weight occur often enough to suggest that differences in the number of units produced, not in the size of the units produced, are mainly responsible for the relative levels of output across industries.

The remaining seven cells in Table 3-7 are labeled Region 1 through Region 7. Each region is associated with a mode of production. The three regions along the diagonal from the top left to the bottom right are designated, respectively, as custom production (Region 1), medium-batch production (Region 3), and mass production (Region 7). These three regions represent the ideal correspondences between the level of cost, the level of output, and the mode of production. If there were not large differences in either the nature of the processing requirements across industries, the unit weight of products, or errors in the estimation of pounds of output, one might expect all industries to fall within the three regions along this diagonal. However, a number of industries fall into Regions 2 and 3, which are adjacent to Region 1, as well as into Regions 5 and 6, which are adjacent to Region 7. Observations in Regions 2 and 3 are designated as custom or small-batch production industries, since they have at least one of the attributes of custom production (high cost or small volume) and none of the attributes of mass production (low cost or large volume). Analogously, observations in Regions 5 and 6 are designated as industries which produce in large batches, since they have at least one of the attributes of mass production and none of the attributes of custom production. The industries which fall into each of these cells will be shown later on in Section 3.6.

The boundary points defining each of the regions are shown along the top and down the left side of Table 3-7, enclosed in parenthesis. For example, the medium-range levels of output are defined as lying within the the closed interval $<m_3, m_4>$. For any industry in Regions 2, 4, or 6, the minimum output is m_3 pounds and the maximum output is m_4 pounds. The medium-range levels of unit cost are defined as lying within the closed interval $<u_3, u_4>$. The maximum and minimum unit cost for any industry in Regions 3, 4, or 5 is u_3 and u_4, respectively. The minimum and maximum values of output for all industries included in the group is given by m_1 and m_6, respectively. The maximum and minimum values of unit cost for all industries included in the group is given by u_1 and u_6, respectively.

The crux of the classification problem is to partition the observations of output and unit cost into low, medium, and high levels. This partitioning is made easier by first subdividing all of the 101 industries in the four-digit data set according to each industry's "parent" major industry group (SIC 34, 35, 36, or 37). While there are no unambiguous boundaries for partitioning the observations, there appears to be rough agreement across all four of the industry groups as to what constitutes low, medium, and high levels of each variable (Table 3-8). For example, within each major group, industries which process nearly 10.0 million pounds of metal per establishment are set off from the "pack" of industries in a scatter plot of value added per pound of metal versus pounds of metal per number of establishments (Figs. 3.6 through 3.9). Similarly, industries which process fewer than 0.33 million pounds of metal per number of establishments are also set off from the majority of the observations. Based on the way in which industries in the scatter plots are distributed across output levels, the following boundaries are suggested for identifying low-, medium-, and high-volume industries.

Table 3-8: Partitioning of Observations of Output and of Unit Cost into Low, Medium, and High Ranges in the Four Major Industry Groups (SIC 34–37)

Major Industry Group	Range of Values		
	Low	Medium	High
	Pounds of Output per Number of Establishments (m/e)		
34	a	0.83–7.9	10.6–35.5
35	0.06–0.28	0.40–5.3	9.6–12.3
36	0.0–0.3	0.50–6.6	9.9–49.8
37	0.0–0.155	0.83–6.8	11.2–49.8
	Value Added per Pound of Metal (va/m)		
34	0.08–0.47	0.64–1.30	1.83–2.27
35	b	0.59–1.40	1.65–13.1
36	0.35–0.55	0.73–1.47	2.52–14.13
37	0.35–0.47	0.70–0.96	2.56–13.1

aMinimum value of m/e is 0.83 in SIC 34
bMinimum value of va/m is 0.54 in SIC 35

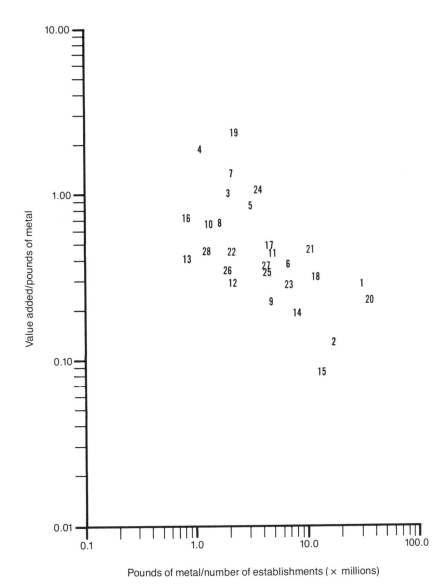

Figure 3-6: Value Added per Pound of Metal versus Pounds of Metal Processed: SIC 34, Fabricated Metal Products (See Table 3-9 for key to figure)

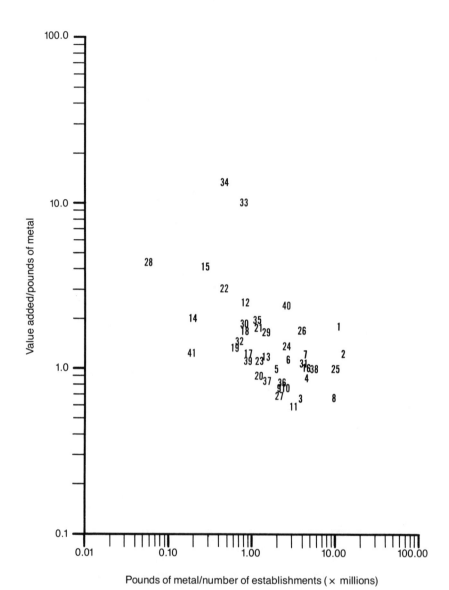

Figure 3-7: Value Added per Pound of Metal versus Pounds of Metal Processed: SIC 35, Machinery, Except Electrical (See Table 3-10 for key to figure)

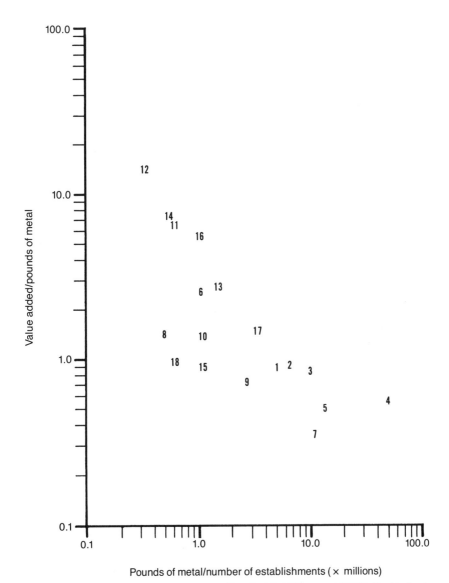

Figure 3-8: Value Added Per Pound of Metal versus Pounds of Metal Processed: SIC 36, Electrical and Electronic Products and Equipment (See Table 3-11 for key to figure)

123

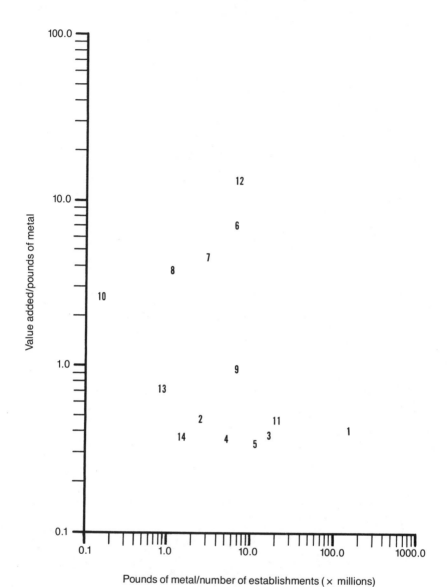

Figure 3-9: Value Added Per Pound of Metal versus Pounds of Metal Processed: SIC 37, Transportation Equipment (See Table 3-12 for key to figure)

Low volume: m/e $\leq$ 0.33
Medium volume: 0.44 $\leq$ m/e $\leq$ 7.9
High volume: m/e $\geq$ 9.6

where

$$m/e = \frac{\text{pounds of metal}}{\text{number of establishments}} \times \text{millions}$$

Within each of the four major groups, a boundary between medium and high levels of unit cost is apparent, since those industries with value added per pound of metal exceeding 1.65 are set off from the "pack" of observations. The boundary between medium- and low-cost industries is drawn at 0.55. Since one of the desired objectives of this classification is to obtain an upper-bound estimate of the proportion of value added that is produced by mass-production industries, a conservative (higher) boundary between low- and medium-cost production is used. If the boundary line between the medium and low level of unit cost were moved downward, fewer industries would be classified as being dominated by mass production. Thus, a large-batch producer might be misclassifed as a mass producer, but it is unlikely that a mass producer would be misclassifed as a large-batch producer.

Based on the way in which industries in the scatter plots are distributed across levels of unit cost, the following boundaries are suggested for identifying industries with low, medium, and high unit cost.

High values of unit cost: va/m $\geq$ 1.65
Medium values of unit cost: 0.59 $\leq$ va/m $\leq$ 1.47
Low values of unit cost: 0.55 $\leq$ va/m

where

$$va/m = \frac{\text{value added}}{\text{pounds of metal processed}}$$

Based on the boundaries specified in Table 3-8, industries are placed in one of the seven regions, as shown in Tables 3-9, 3-10, 3-11, and 3-12. In figures 3-6, 3-7, 3-8, and 3-9, each four-digit industry is designated by an integer (1, 2, 3, 4, . . .). A list identifying the SIC code and the name of the industry designated by each integer is given in Tables 3-9, 3-10, 3-11, and 3-12. These lists are organized into the seven regions described earlier, so an industry's position in the figure will roughly correspond to its position in the list. In the list, industries within a region are ordered according to their measure of value added per pound of metal. The industry with the highest measure of unit cost is listed first. Following the results for the separate groups, the results are shown for the pooled data set (Figure 3-10). A separate list showing the identity of all of the industries in Figure 3-10 is not necessary, since it would be the same as combining all of the lists for SIC 34, 35, 36, and 37. The pooled four-digit data set by average unit cost and by output is also shown (Table 3-13).

Table 3-9: Clustering of Industries by Mode of Production: SIC 34

	Pounds of Metal (in millions) per Number of Establishments						
	(0.0) Low (0.0)	(0.83)	Medium	(7.9)	(10.6)	High	(35.5)
	Region 1		Region 2				
(2.27) High			(19) 3463 Nonferrous forgings				
(1.83)			(4) 3425 Hand saws and saw blades				
	Region 3		Region 4			Region 5	
(1.3) Medium (0.64)			(7) 3432 Plumbing fittings, brass goods (24) 3494 Valves and pipe fittings (3) 3423 Hand and edge tools, nec (5) 3429 Hardware, nec (16) 3451 Screw machine products (8) 3433 Heating equipment, except electrical (10) 3442 Metal doors, sash, and trim				
			Region 6			Region 7	
(0.47) Low (0.08)			(17) 3452 Bolts, nuts, rivets, washers (11) 3443 Fabricated platework, boiler shops (28) 3499 Fabricated metal products, nec (22) 3469 Metal stampings, nec (13) 3446 Architectural metalwork (6) 3431 Metal sanitary wares (27) 3498 Fabricated pipe and fittings (25) 3495 Wire springs (26) 3496 Miscellaneous fabricated wire products (12) 3444 Sheet metalwork (23) 3493 Steel springs, except wire (9) 3441 Fabricated structural metal products (14) 3448 Prefabricated metal buildings			(21) 3466 Crowns and closures (18) 3462 Iron/steel forgings (1) 3411 Metal cans (20) 3465 Auto stampings (2) 3412 Metal barrels, containers (15) 3449 Miscellaneous metalwork	

Value Added per Pound of Metal

nec = not elsewhere classified in census documents

3.6.4. THE POOLED FOUR-DIGIT SIC DATA SET

The industries located in the lowest and highest values of unit cost and output are shown in Table 3-13. The 10 industries with the lowest levels of unit cost are all in SIC 34, Fabricated Metal Products. For the most part, these 10 industries shape and/or join inexpensive types of steel. None of these industries purchase considerable amounts of processed metal products. The five industries with the highest unit cost are:

3662 radio and TV communication equipment
3574 calculating and accounting equipment
3761 guided missiles and space vehicles
3573 electronic computing equipment
3676 electronic resistors

Four of these industries also have the highest ratio of processed metal costs to

Table 3-10: Clustering of Industries by Mode of Production: SIC 35

	Pounds of Metal (in millions) per Number of Establishments		
	(0.06) Low (0.28)	(0.44) Medium (5.3)	(9.6) High (12.3)
(13.07)	Region 1	Region 2	
High		(34) 3574 Calculating and accounting machines	
		(33) 3573 Electronic Computing Equipment	
	(28) 3565 Industrial patterns		
	(15) 3545 Machine tool accessories	(22) 3555 Printing trades machinery	
		(12) 3541 Machine tools, metalcutting	
		(40) 3592 Carburetors, piston rings, etc.	
	(14) 3544 Special tools, dies, jigs, etc.	(35) 3576 Scales and balances	
		(21) 3554 Paper industry machinery	
		(30) 3567 Industrial furnaces and ovens	
		(18) 3551 Food products machinery	
		(26) 3563 Air and gas compressors	(1) 3511 Turbines, generator sets
(1.65)		(29) 3566 Speed changers, drives, gears	
(1.40)	Region 3	Region 4	Region 5
		(32) 3569 General industrial machinery, nec	
		(19) 3552 Textile machinery	
	(41) 3599 Machinery, nec	(24) 3561 Pumps and pumping equipment	
		(7) 3533 Oilfield machinery	
		(17) 3549 Metalworking machinery, nec	
		(13) 3542 Machine tools, metal forming	
		(39) 3589 Service industry machinery, nec	
		(6) 3532 Mining machinery	
		(23) 3559 Special industry machinery, nec	
		(31) 3568 Power transmission equipment	
		(16) 3547 Rolling mill machinery	
		(38) 3585 Refrigeration and heating equipment	
Medium		(5) 3534 Elevators and moving stairways	
		(4) 3524 Lawn and garden machinery	
		(20) 3553 Woodworking machinery	
		(37) 3582 Commercial laundry equipment	
		(36) 3581 Automatic merchandising machines	
		(10) 3536 Hoists, cranes, monorails	(2) 3519 Internal combustion engines, nec
		(9) 3535 Conveyors, conveying equipment	(25) 3562 Ball and roller bearings
		(27) 3564 Blowers and fans	(8) 3531 Construction equipment
		(3) 3523 Farm machinery and equipment	
(0.59)		(11) 3537 Industrial trucks and tractors	
Low		Region 6	Region 7

(Left vertical axis label: Value Added per Pound of Metal)

nec = not elsewhere classified in census documents

Table 3-11: Clustering of Industries by Mode of Production: SIC 36

Value Added per Pound of Metal	Pounds of Metal (in millions) per Number of Establishments								
	(0.0)	Low	(0.33)	(0.49)	Medium	(6.6)	(9.9)	High	(49.8)
		Region 1			Region 2				
(14.13) High		(12) 3662 Radio and T.V. communication equipment							
					(14) 3676 Electronic resistors				
					(11) 3651 Radio and TV receiving sets				
					(16) 3678 Electronic connectors				
					(13) 3675 Electronic capacitors				
(2.52)					(6) 3643 Current-carrying wiring devices				
		Region 3			Region 4			Region 5	
(1.47) Medium					(17) 3694 Electrical equipment for internal combustion engines				
					(8) 3645 Residential lighting fixtures				
					(10) 3648 Lighting fixtures, nec				
					(18) 3699 Electronic equipment and supplies, nec				
					(2) 3621 Motors and generators				
					(15) 3677 Electronic coils, transformers, inductors				
					(1) 3612 Transformers				
(0.74)					(9) 3646 Commercial lighting fixtures			(3) 3631 Household cooking equipment	
		Region 6						Region 7	
(0.54) Low								(4) 3633 Household laundry equipment	
								(5) 3639 Household appliances, nec	
(0.35)								(7) 3644 Noncurrent-carrying wiring devices	

nec = not elsewhere classified in census documents

basic metal costs, which indicates that they are mostly involved in the assembly of components. This suggests that unit cost is strongly influenced by the nature of the processing requirements. This is explored in greater detail in Chapter 4.

The five industries which process the most pounds of metal per establishment are:

3711 motor vehicles and car bodies
3633 household laundry equipment
3465 automobile stampings
3411 metal cans
3743 railroad equipment

The five industries which process the smallest amount of metal per establishment are:

Table 3-12: Clustering of Industries by Mode of Production: SIC 37

Value Added per Pound of Metal	Pounds of Metal (in millions) per Number of Establishments								
	(0.0)	Low	(0.155)	(0.84)	Medium	(6.8)	(11.2)	High	(149.05)
(13.10)		Region 1			Region 2				
					(12) 3761 Guided missiles and space vehicles				
					(6) 3721 Aircraft and parts				
High					(7) 3724 Aircraft engines and parts				
					(8) 3728 Aircraft parts and auxiliary equipment				
(2.56)		(10) 3782 Boat building and repair							
		Region 3			Region 4			Region 5	
(0.96)					(9) 3731 Ship building and repair				
Medium									
(0.70)					(13) 3792 Travel trailers and campers				
		Region 6			Region 7				
(0.47)					(2) 3713 Truck and bus bodies				
								(11) 3743 Railroad equipment	
					(14) 3799 Transportation equipment, nec			(1) 3711 Motor vehicles and car bodies	
Low					(4) 3715 Truck trailers			(3) 3714 Motor vehicle parts, accessories	
(0.35)								(5) 3716 Motor homes	

nec = not elsewhere classified in census documents

3565 industrial patterns
3732 boat building and repair
3599 machine shops
3544 special tools and dies
3545 machine tool accessories

It is important to note that, with the exception of railroad equipment, the industries which make the largest, heaviest products, such as boilers, turbines, and structural metal for buildings, are not among the five industries with the largest output per establishment. Also, industries which produce some of the smallest, lightest products, such as screws, bolts and nuts, and electronic components, are not among the five industries with the smallest output per establishment. While there are several notable cases where the large unit weight of an industry's product probably resulted in the industry being misclassified (most notably with engines and turbines in SIC 35), it should be evident that differences in unit weights among products do not dominate the whole analysis.[16]

16. In the engine and turbine industry (point 1 in SIC 35), it appears that the large unit weight of the product results in a misclassification. For purposes of allocating value added by mode of production, the engine and turbine industry is included in custom-small batch production. In SIC 35, the output of the ball and roller bearing industry (point 25) is roughly the same as that of the engine and

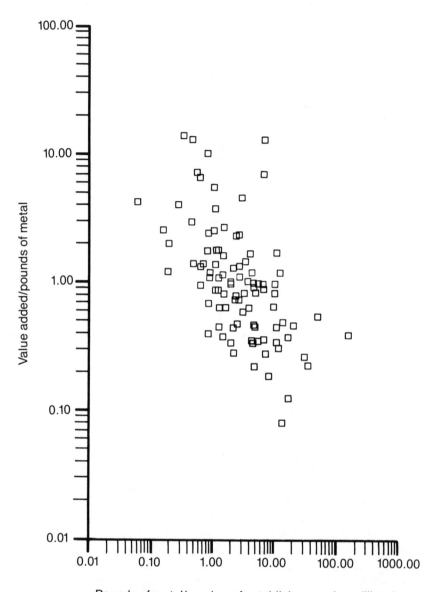

Figure 3-10: Value Added per Pound of Metal versus Pounds of Metal Processed Pooled Four-Digit Data Set

Table 3-13: Average Cost and Output for Pooled Four-Digit Data Set, SIC 34–37

Middle of Interval	Number of Observations	Industries (by SIC Codes)
	Value Added Per Pound of Metal (in Natural Log Units)	
-2.500	1	3449
-2.250	0	
-2.000	1	3412
-1.750	1	3448
-1.500	2	3441, 3465
-1.250	4	3411, 3493, 3444, 3462
-1.000	11	3496, 3716, 3495, 3644, 3498, 3715, 3431, 3799, 3714, 3711, 3446
-0.750	8	
-0.500	6	
-0.250	12	
0.000	15	
0.250	11	
0.500	9	
0.750	3	3544, 3463, 3592
1.000	5	3541, 3643, 3732, 3675, 3555
1.250	1	3728
1.500	3	3545, 3565, 3724
1.750	1	3678
2.000	3	3651, 3721, 3676
2.250	1	3573
2.500	2	3761, 3574
2.750	1	3662
	Pounds of Metal Processed Per Number of Establishments (in Natural Log Units)	
-2.750	1	3565
-2.500	0	
-2.250	0	
-2.000	0	
-1.750	3	3732, 3599, 3544
-1.500	0	
-1.250	1	3545
-1.000	1	3662
-0.750	3	3555, 3574, 3645
-0.500	4	3676, 3651, 3699, 3552
-0.250	9	3569, 3573, 3567, 3551, 3541, 3451, 3446, 3792, 3589
0.000	7	
0.250	6	
0.500	6	
0.750	10	
1.000	7	
1.250	6	
1.500	11	
1.750	2	
2.000	6	
2.250	5	3531, 3631, 3562, 3644, 3466
2.500	6	3511, 3716, 3462, 3519, 3449, 3639
2.750	2	3714, 3412
3.000	1	3743
3.250	0	
3.500	2	3411, 3465
3.750	0	
4.000	1	3633
4.250	0	
4.500	0	
4.750	0	
5.000	1	3711

While sources of noise in this classification procedure are potentially troublesome, it is still believed that most of the industries are correctly classified. For example, there is no doubt that the auto-related industries—automotive stampings, motor vehicles and car bodies, and motor vehicle parts and accessories—are all dominated by mass producers, and that industries that produce aircraft, computers, metalcutting machine tools, and special tools and dies are all dominated by custom or small-batch producers. It appears that most of those industries where the dominant mode of production is known are correctly classified, and that in most other cases the classifications are not inconsistent with what is generally known about which types of products are usually custom-, batch-, and mass-produced.

Of special importance is the belief that the analysis identifies all of the industries in the sample dominated by mass production. The boundaries between regions were established to err on the side of locating some industries which are probably large-batch producers in the mass-production region, as opposed to the other way around. For example, in SIC 34, crowns and closures (SIC 3466) and iron and steel forgings (SIC 3462) fall within Region 7 and are classified as mass-production industries, even though these are believed to be dominated by large-batch production. In SIC 37, locomotives (SIC 3743) and motor homes (SIC 3716) are classified as mass-production industries, even though they are believed to be dominated by large-batch production. However, if the boundaries between medium and high levels of unit cost and output were to be reestablished so as to place these industries in the region of large-batch production, it would not be possible to have the same boundary points apply across all four industry groups. The estimated distribution of value added by the mode of production derived from this analysis can be regarded as giving an upper bound on the proportion of value added coming from mass-production industries and, correspondingly, as giving a lower bound on the proportion coming from batch-production industries.

3.7. The Distribution of Value Added by Mode of Production

The distribution of value added is estimated in Table 3-14, based on the classification of four-digit industries in each of the four major metalworking groups. The value added within each of the regions is also summed across the four major groups to estimate the distribution by mode of production for metalworking as a

---

turbine industry. This is just one of many examples in this data set where industries producing products which are small in size have a larger output than industries producing products which are large in size. While differences in the unit weight of products undoubtedly introduces some noise into the analysis, it is clearly *not the case* that differences in the unit weight of products are primarily responsible for the differences in the levels of output across industries.

Table 3-14: Percentage Distribution of Value Added by Mode of Production for Industries in Sample, SIC 34–37

Region and Mode of Production	34	35	36	37	Total, 34–37
R1 + R2 + R3 (custom and small batch)	1.1	41.0	58.5	30.6	31.5
R4 (medium batch)	28.3	42.5	30.2	6.9	26.4
R5 + R6 (large batch)	45.5	16.5	3.1	2.9	16.4
R7 (mass)	25.1	0.0	8.2	59.6	25.7
Sample coverage of value added for all industries	94.2	95.8	51.2	97.4	86.0

whole (SIC 34–37). Since industries with low material coverage within each major group are omitted, the combined value added for industries in the sample is less than the combined value added for all industries in each major group. The coverage of the sample is given by the ratio of the combined value added for the industries in the sample to the combined value added for all industries in the major group. The industries included in the sample account for practically all of the value added in major groups SIC 34, 35, and 37, but for only half of the value added and output in SIC 36. Across the four groups, the industries in the sample account for 86 percent of the total value added for all of the industries in SIC 34–37.

The seven regions are grouped together to form four regions in Table 3-14, since noise in the data introduces the possibility of misclassification and since there is some degree of arbitrariness in specifying the boundaries between regions. Observations in Regions 1, 2, and 3 are classified as custom-small-batch production industries. Observations in Region 4 are classified as medium-batch production industries. Observations in Regions 5 and 6 are classified as large-batch production industries. Observations in Region 7 are classified as mass-production industries. Construing batch production broadly to mean everything that is not mass production, Regions 1–6 would cover the full spectrum of batch sizes, ranging from very small batch (custom) to large-batch production. The industries in Regions 1–6 account for just under 75 percent of the value added in the sample. Stated another way, the industries with the lowest levels of unit cost and with the highest levels of output, which are assumed to be the mass-production industries, account for about 25 percent of the value added in the sample. The proportions accounted for by mass production might be slightly understated, since several industries in SIC 36 which manufacture household appliances and which are probably dominated by mass production are excluded from the sample. This group of excluded industries is composed of household refrigerators and freezers (SIC 3632), electric housewares and fans (SIC 3634), household vacuum cleaners (SIC 3635), sewing machines (3636), and telephone and telegraph apparatus (3661). Even if these industries were included in Region 7, mass production would still account for only 28 percent of the value added in the sample (and for only 25 percent of total value added in SIC 34–37, including all industries). An upper limit on the proportion of value added and of output accounted for by high-volume production, including both large-batch and mass production, is derived by considering all of the industries in Regions 5, 6, and 7 to be high-volume producers. The industries in these regions account for just over 42 percent of the value added in the sample. This analysis corroborates the claim that most of the value added in the metalworking sector is accounted for by products which are batch produced.

The distribution of the total value of output by mode of production is given in the top half of Table 3-15. Total value of output includes material cost as well as

Table 3-15: Percentage Distribution of Value of Output and of Employment by Mode of Production for Industries in Sample, SIC 34–37

Region and Mode of Production	34	35	36	37	Total 34–37
			Total Value of Output		
R1 + R2 + R3 (custom and small batch)	1.0	36.3	55.1	20.1	24.8
R4 (medium batch)	25.3	44.3	31.4	4.9	23.3
R5 + R6 (large batch)	45.1	19.4	3.8	2.8	16.2
R7 (mass)	28.6	0.0	9.7	72.2	35.7
Sample coverage of output in all industries	95.1	96.1	51.0	98.0	88.0
			Production Worker Employment		
R1 + R2 + R3 (custom and small batch)	0.9	41.8	51.5	24.9	27.8
R4 (medium batch)	27.6	42.1	37.9	12.9	29.4
R5 + R6 (large batch)	51.9	16.1	3.1	4.6	20.2
R7 (mass)	19.6	0.0	7.5	57.6	22.6
Sample coverage of production worker employment in all industries	94.5	95.9	52.5	97.3	85.8

value added. The mass-production industries account for a higher percentage of the total value of output in the sample than of value added because they process more material per dollar of value added than the batch-production industries. Nonetheless, the batch-production industries (those in Regions 1-6) still account for almost two-thirds of the total value of output in the sample, compared to three-fourths of the value added.

The distribution of the number of production workers by mode of production is also shown in Table 3-15. The mass-production industries account for a smaller percentage of the production workers because they are relatively less labor intensive than the batch-production industries. The batch-production industries (again, Regions 1-6) account for almost 78 percent of the employment of production workers, compared to 75 percent of the value added.

4

Potential Impacts on Production Cost

The extent to which production cost could be decreased as a result of using robots is examined in more detail in this chapter. In the first case considered, robots replace a fraction of the production work force in a factory without affecting throughput. Even if Level I and Level II robots were fully utilized and 10–30 percent of the workers were eliminated, the percentage of reduction in unit cost resulting from labor savings would be relatively modest. In the second case considered, it is assumed that the factory throughput is modestly increased as a result of robots operating more continuously than human operators and a reduction of scrap through tighter control over operating conditions. Because of these improvements in efficiency, capacity is increased by 10–20 percent without increasing the number of shifts or days of operation. Capital costs required to achieve this improvement *are not* considered, so the calculated reductions in unit cost derived for this case are only rough approximations which should be viewed as upper-bound estimates. With this relatively modest increase in throughput, the reduction in unit cost would be several times greater than if only labor cost were reduced. Most of the decrease in unit cost is due to the increase in throughput. Eliminating a larger percentage of production labor cost contributes only marginally to the additional reduction in unit cost. This case focuses attention on the potential benefits of using robots to increase the capacity of a factory. It also serves to motivate the third case, which is a more comprehensive analysis of the potential for increasing output of a factory and for reducing unit cost.

In the third case, it is assumed that a factory organized to produce batches of specialized products is reorganized to integrate robots with other types of computer-aided manufacturing (CAM) technology. As a result of the reorganization and integration, the factory can produce specialized products in a continuous fashion around the clock, and the level of output is comparable to that of a factory mass producing a standardized product. The potential for increasing the level of output is derived from estimates of the percentage of time machines are currently utilized in conventionally organized factories producing low and medium volumes of specialized products. Estimates of the resulting decrease in unit cost (including changes in capital cost) are derived from the analysis of cost variations with the level of output across metalworking industries presented in Chapter 3. (The Chapter 3 analysis is substantially expanded here for this purpose.)

This reorganization scenario examines what would happen to unit cost in a batch-production factory if it were to have a level of output that is comparable to a factory mass producing a standardized product. While robotic manipulators would be necessary for such a factory, they would compose only a relatively small part of the hardware and total capital cost. Thus, the reorganization scenario examined in the third case is more appropriately viewed as an analysis of the potential impact on unit cost when fully utilizing *flexible automation,* including robotics, as opposed to an analysis of the isolated impact of robotics.

4.1. Case 1. Robot Use Does Not Affect Throughput

This case assumes that robots are retrofitted into an existing factory which is designed and organized for human workers. While the use of robots reduces labor cost as a result of eliminating some jobs, there is no effect on the quantity of output produced since the installed robots are essentially islands of automation. Even if the capacity of the work stations using robots increases, there are many other bottlenecks in the plant, so there is no increase in the overall capacity of the factory.

The percentage of decrease in the cost of output for Case 1 is calculated from the model presented in Chapter 2. The cost reductions are calculated under the assumption that robots replace a given percentage of production workers and that there is no increase in throughput. The results in Table 4-1 give an upper

Table 4-1: Percentage of Reduction in Production Cost Resulting from Eliminating a Fraction of Production Labor Cost with No Change in Throughput

Decrease in Production Labor (ΔL)	Production Labor Share of Costs (v_L)			
	10%	20%	30%	40%
1%	0.1	0.2	0.3	0.4
5%	0.5	1.0	1.5	2.0
10%	1.0	2.0	3.0	4.0
20%	2.0	4.0	6.0	8.0
30%	3.0	6.0	9.0	12.0

See Chapter 2 for an explanation of the model used to calculate reductions in production cost.

bound on the decrease in cost that would result from a given percentage of decrease in production labor requirements. The data represent upper bounds because they do not include increases in capital cost or the likely increases in salaried labor cost. If the capital investment can be recovered from the cost savings in a relatively short period of time (relative to the horizon considered), then one can realistically ignore capital cost. (See Chapter 2 for a discussion of the cost of installing robots and calculation of payback periods.) The decrease in cost depends on the reduction in production labor requirements, as well as on the production labor share of total cost. Table 4-2 shows that the production labor share of total cost (including materials cost) ranges from 13 to 23 percent for most of the 101 metalworking industries included in this analysis. The average production labor cost share is 18.7 percent. To facilitate the discussion of Table 4-1, attention is focused on the cost reduction realized when production labor cost constitutes 20 percent of total cost.

If only 1 percent of the workers in a firm or an industry were to be displaced by 1990, as is implied by the Hunt and Hunt (1983) study and other market forecasts (see Chapter 2), then the decrease in production cost would be negligible (0.2 percent) for most industries. If Level I robots were to be fully utilized and 10 percent of the production workers were eliminated, the decrease in production cost would be 2 percent in most industries, which is still only a modest decline. If 30 percent of the production workers were to be eliminated (full utilization of Level II robots), then the decrease in production cost would be on the order of 6 percent in most industries. While a 6 percent decrease is not insignificant, it

Table 4-2: Production Labor Share of Total Output for Metalworking Industries in Sample

Middle of Interval	Number of Observations
0.0800	1
0.1000	4
0.1200	5
0.1400	12
0.1600	23
0.1800	20
0.2000	10
0.2200	10
0.2400	5
0.2600	4
0.2800	3
0.3400	2
0.3600	1
0.4200	1
Average =	0.18657

Source: Compiled from the 1977 Census of Manufactures (Bureau of the Census, 1981c,d).

Production labor share = $\frac{\text{production labor cost}}{\text{total output}}$

Total output = production labor cost + salary labor cost + nonlabor value added + materials cost does not represent a dramatic cost reduction either. The conclusion from Case 1 is that if the primary effect of robots is only to decrease production labor requirements, leaving the quantity of output unchanged, then the substitution of robots for workers will not have a subtantial impact on total production cost.

4.2. Case 2. Robot Use Moderately Increases Throughput

Case 2 results show that if output were to increase by only a relatively modest amount, the decrease in production cost would be substantially greater than if only labor cost were reduced. What level of increase is considered a "modest" increase in output? To help answer this question, several examples of how the use of robots has actually increased the rate of throughput in factories are shown in Table 4-3. Most of the information in the table comes from interviews with applications engineers in several factories using robots. The senior applications engineer from Manufacturer A reports that as a rule of thumb in material handling and machine loading applications, a robot will achieve a 10–12 percent increase in throughput over a human operator, since it works consistently without taking breaks. (Manufacturer A first installed robots in 1965 and had 100 robots in use as of mid-1982.) The examples provided by Manufacturer A also show that within a particular work station the impacts of robots on throughput might be smaller or substantially larger than this rule-of-thumb estimate.

In Manufacturer B's process, the way in which the machine is operated affects yield of acceptable parts produced, so that robot use affects product yield as

Table 4-3: Examples of the Impacts of Retrofitting Robots on Labor Requirements and Throughput

Application	Labor Change Per Shift	Throughput Change Per Shift
Manufacturer A: Metalcutting, Metalforming, and Assembly		
Heat treating	from 1 to 1/3	12% increase
Material handling in forging press	from 2 to 1/2	from 150/hour to 400/hour
Arc welding	no change	from 30% arctime (maximum) to 90% arctime (average)
Palletizing in a machining center	from 2 to 0	10–12% increase
Machine loading/unloading	from 2 to 1	5% increase; tool-change time limits increase
Assembly machine ("hard automation")	from 5 to 3	from 16 assemblies/hour to 35 assemblies/hour
Manufacturer B: Die Casting		
Die casting machine operation	1 to 0	9% increase in utilization and 10–15% increase in yield; 20–25% total increase in throughput

well as machine utilization. The senior applications engineer from Manufacturer B has reported:

> Because automatic operation is continuous, i.e., no breaks or lunch periods, greater utilization of casting machines is realized. Experience has shown that with a given group of machines in the automatic mode, a gain of 9 percent in productive capacity is obtained compared with the output from the same group of machines using operators. Again, because of continuous operation and elimination of the variables attributed to operators, automatic operation will reduce rejects from 10 to 15 percent over that experienced using operators. This added yield of good pieces results from dimensional as well as cosmetic factors. (Marshall, 1975: 5)

Taken together, a 9 percent increase in the amount of time the machine is operating and a 10–15 percent increase in the number of acceptable pieces produced results in a 20–25 percent increase in the number of acceptable parts manufactured. Based on these two examples, it is assumed that a 10–20 percent increase in throughput can be realized in machine loading and material handling applications as a result of the robot working continuously during a shift and as a result of reducing material waste through stricter control over operating conditions. If the factory were to operate at this higher rate of throughput over an entire period (such as a shift, a month, a year, etc.), it would result in a 10–20 percent increase in the level of output produced per period. This increase could potentially be realized without increasing the number of shifts worked per day and the number of days worked per year. This percentage range represents the amount by which the level of output could be increased without having to reorganize the factory substantially so that it could run unattended during those periods when it was previously unscheduled for operation.

The decrease in production cost that would result if throughput were increased by either 10 or 20 percent for various production labor reductions is shown in Table 4-4. Increases in capital cost and salaried labor cost required to install the robots and increases in material cost required to increase throughput are not considered. Because of these omissions, and because it is assumed that overall factory throughput increases by the same amount as the increase in throughput of robotic work stations, the reductions in total production cost shown in Table 4-4 should be viewed as upper-bound estimates. For a given reduction in production labor requirements, the decrease in unit cost is substantially larger than in the previous case where there is no accompanying increase in throughput. Again, in order to facilitate discussion, attention is focused on the cost reduction realized when production labor cost constitutes 20 percent of the total cost. With a 10 percent increase in throughput, unit cost would decrease by 9 percent, with production labor cost held constant. If in addition to the increase in throughput, labor inputs are reduced by 10 or 30 percent, unit cost will decrease by 11 percent and 14.5 percent, respectively. With a 20 percent increase in throughput, unit cost will decrease by 16.7 percent, with labor cost

Table 4-4: Percentage of Reduction in Production Cost Resulting from 10 Percent and 20 Percent Increases in Throughput and from Eliminating a Fraction of Production Labor Cost

Decrease in Production Labor (ΔL)	Production Labor Share of Costs (v_L)			
	10%	20%	30%	40%
	10% Increase in Output			
0%	9.1	9.1	9.1	9.1
5%	9.5	10.0	10.4	10.9
10%	10.0	10.9	11.8	12.7
20%	10.9	12.7	14.5	16.4
30%	11.8	14.5	17.3	20.0
	20% Increase in Output			
0%	16.7	16.7	16.7	16.7
5%	17.1	17.5	18.3	19.2
10%	17.5	18.3	20.0	21.7
20%	17.9	19.2	21.7	24.2
30%	18.3	20.0	23.3	26.7

See Chapter 2 for an explanation of the model used to calculate reductions in production cost.

held constant. This is almost three times the decrease in unit cost that would be realized in most industries if 30 percent of production laborers were eliminated while maintaining a constant throughput (Table 4-1). Reducing labor input by 10 or 30 percent in addition to the throughput increase results in a cost reduction of 18.3 and 20 percent, respectively.

The magnitude of the results shows that if robot use were to result in an increase in the quantity of output, in addition to a reduction in labor requirements, then the reduction in unit cost would be substantially greater than if only labor requirements were reduced. Most of the decrease in cost is due to the increase in output, and the marginal benefits of eliminating a fraction of production labor cost are small. This example suggests that it is important to analyze in further detail the effects on unit cost in batch-production facilities if the level of output is increased. The discussion proceeds with estimates of the potential for increasing output in a batch-production facility if it were utilized around the clock for most of the year. Following that, estimates are made of the reductions in unit cost and labor requirements that would be achieved if output were to increase by these large amounts.

4.3. Case 3. Robotics and CAM Dramatically Increase the Factory's Maximum Level of Output

The key feature of the third case is that the use of robots, in conjunction with of other types of automation (e.g., CAM systems), makes it possible to increase the capacity of the factory substantially without proportionately increasing labor and capital inputs. As a result, the average cost per unit decreases as the quantity of output increases. The primary reason for the increase in capacity is that the reorganized factory is more fully utilized than a conventional facility. It is designed to stay in operation around the clock for the whole year. In contrast, many conventionally organized batch production factories do not operate on weekends or holidays, and they may also close down during parts of the days they are scheduled for operation. The secondary reason for the increase in capacity is that the throughput rates are higher in the reorganized factory than in the conventional one as a result of robots operating more continuously and stricter control over operating conditions.

4.4. The Potential for Increasing Output

To realize the full potential of its capital equipment, a plant would have to be operated around the clock every day of the year, with allowance for preventive maintenance. I have estimated the amount by which output could be increased if plants were fully utilized, based on the breakdowns of theoretical capacity for typical low-, medium-, and high-volume manufacturing establishments published by the Machine Tool Task Force (Mayer and Lee, 1980a,b). These breakdowns were already presented in Chapter 3 and are analyzed here in more detail. A second estimate of the potential for increasing output is also made based on the measure of machine utilization introduced in Chapter 3.

The breakdowns of theoretical capacity for typical types of low-, medium-, and high-volume establishments can be reorganized into three categories:

1. time not scheduled for production (holidays, plant shutdowns, and incomplete use of the second and third shifts);
2. time lost as a result of inefficiencies (equipment failure, tool changes, loading/unloading, setup, etc.); and
3. time actually used to process materials.

First, I have calculated how much more productive time would be available during a year if the time normally not scheduled for production could be utilized. For each type of plant, the number of days per year that it is normally open for operation and the number of hours per day that it is normally scheduled to produce are given in Table 4-5. Notice that a fourth type of plant, designated as the low-volume two-shift operation, has been added to the table in order to include the case where a plant operates for two full shifts.[1] It is assumed that one

---

1. In the representative plants described by the Machine Tool Task Force, the high-volume establishment operates on 3 shifts, the medium-volume establishment has 1.3 shifts, and the low-volume plant has 1 shift. I have added the case of a low-volume plant that typically operates on two shifts. Rather than describing a level of output below that of the typical medium-volume plant, the low-volume, two-shift case is so named because it describes the low-volume plant but with the modification that it operates for two full shifts per day instead of one. The low-volume plant operating 16

Table 4-5: Estimated Planned Production Time in Low-, Medium-, and High-Volume Metal Fabricating Manufacturing

	High Volume	Low Volume (two-shift operation)[a]	Medium Volume	Low Volume (one-shift operation)
Maximum days per year available	365	365	365	365
Days per year open for operation	286	241	263	241
Hours per day allotted for:				
Production	22.2	16.0	10.7	8.0
Preventive maintenance[b]	1.0	1.0	1.0	1.0
Unscheduled (idle) time	0.8	7.0	12.3	15.0
Scheduled production time proportion of maximum available time	72.6%	44.0%	32.0%	22.0%

Source: Derived from breakdowns of theoretical capacity given in Mayer and Lee, 1980a.

[a]Based on the breakdown of theoretical capacity for low-volume manufacturing, but assumes that the plant operates on two shifts instead of one.

[b]Most factories do not stop production on a daily basis to perform scheduled preventive maintenance. Machines are typically serviced on an as-needed basis. Assuming one hour per day of scheduled maintenance may be a high estimate. Major machine overhauls and repairs are typically carried out during scheduled plant shutdowns.

hour per day is required for preventive maintenance, so the maximum number of hours is based on a 23-hour day.

For each of the four types of plants, the baseline figure for the amount of productive time is given by

productive time baseline = dh

where

d = days per year open for operation

h = hours per day scheduled for operation

If the plant were to produce for 23 hours during those days that it is normally scheduled to be shut down, the percentage of increase in available productive time would be given by

$$\left[\frac{23(365 - d)}{dh} \right] 100 \qquad (4.1)$$

If the plant were to produce for a full 23 hours during those days that it is open for production, the percentage of increase in available productive time would be given by

hours per day is in use more time than the so-called medium-volume plant, which is assumed to operate only 10.7 hours per day. Thus, the low-volume, two-shift case more appropriately represents a type of plant whose level of output falls between that of typical high-volume and medium-volume plants.

$$\left[\frac{(23 - h)d}{dh}\right] 100 \qquad\qquad (4.2)$$

The total percentage of increase in time available for production is given by the sum of Equations 4.1 and 4.2. It is assumed that in each type of plant, each 1 percent increase in the amount of productive time available would result in a 1 percent increase in the level of output. In other words, the plant could sell all that it could produce.

The amounts by which time available for production could be increased in each of the four types of plants is shown in Table 4-6. The calculations show that plants already operating on three shifts (assumed to be high-volume plants) could increase their output by 30 percent, mostly as a result of recouping the time lost during plant shutdowns. The potential increase in output for the low- and medium-volume plants, which are assumed to operate on fewer than three shifts and to be closed more days of the year, is substantially greater. Output could be increased by nearly 120 percent in plants operating on 2 shifts, by nearly 200 percent in plants operating on 1.3 shifts, and by over 330 percent in plants operating on only a single shift.

I have already stated that, in addition to extending the amount of time available for production per year, the use of robots can increase throughput without an increase in the amount of time worked, because they can work more efficiently than human operators during a given time period. Based on interviews with experienced robot users, I have stated that the level of throughput can sometimes be increased by 10–20 percent in particular work stations using robots. In the following discussion, I make another estimate of the increase in throughput that could be realized as a result of working more efficiently during a given time period, based on the breakdowns of theoretical capacity in low-, medium-, and high-volume metalworking plants made by the Machine Tool Task Force in 1980 (Mayer and Lee, 1980a).

Estimated potential increases in throughput that could be achieved from recouping scheduled operation time not used for direct processing of materials

Table 4-6: Potential Percentage Increases in Available Production Time

	Increase in Utilization of Days Plant Is Closed	Increase in Utilization of Unscheduled Production Time	Total Percentage Increase in Output
High volume	28	3	31
Low volume (two-shift operation)	74	43	117
Medium volume	83	115	198
Low volume (one-shift operation)	148	187	335

Source: Derived from estimates of available time in Table 4-5.

Table 4-7: Potential Increases in Throughput from Working More Efficiently During Scheduled Production Time: High-Volume Manufacturing

Function	Operating Time[a] (%)	Robots Only		Robots + CAM	
		Potential Reduction (%)	Adjusted Operating Time (%)	Adjusted Potential Reduction (%)	Adjusted Operating Time (%)
Load/unload, noncutting[b]	20	10	18	25	15
Work station allowances	20	40	12	80	4
Inadequate storage	10	0	10	50	5
Tool change[b]	10	0	10	20	8
Equipment failure	10	0	10	0	10
Direct processing of materials	30	0	30	0	30
Total[c]	100		90		72
Potential throughput index	1.0		1.11		1.39

[a]Breakdown of theoretical capacity from Dallas, 1981; and Mayer and Lee, 1980a, b.
[b]Already highly automated in high-volume plants.
[c]Total scheduled production time.

(idle time, loading/unloading, setup, gauging, etc.) are shown in Tables 4-7 through 4-9. The estimated improvements are based on my informed judgment, and have been reviewed by several industry experts. Two levels of improvement are distinguished: increases in throughput achieved (1) by robots only and (2) by more fully integrating robots with other forms of CAM technologies.[2] Much of the time not spent on direct material processing is taken up by machine-related functions—equipment limitations, tool changing, and equipment failures. However, a sizable fraction of work time typically spent processing materials is lost because of management and work-force practices, including personal time breaks, late starts, early quits, material handling, excessive machine adjustments, and in-line storage losses due to scheduling inefficiencies. Personal time, late starts, and early quits, and some fraction of the material-handling time are expected to be nearly eliminated by replacing workers with robots. Time losses due to tool changing, equipment failures, excessive machine adjustments, setups, and scheduling inefficiencies will probably not be affected directly by robots, but might be reduced if more aspects of factory work

2. An example of integrating robots with other CAM techniques is the use of *flexible manufacturing systems* for machining operations instead of stand-alone metalcutting machines. Integrating the functions of several individual machines into one system would substantially reduce the number of setups and the amount of tool changing, loading and unloading, and time the part spends in transit between machines. Thus, the potential increases in output are assumed to be substantially larger for the case of robots used in conjunction with other CAM technologies.

Table 4-8: Potential Increases in Throughput from Working More Efficiently During Scheduled Production time: Medium-Volume Manufacturing

		Robots Only		Robots + CAM	
Function	Operating Time[a] (%)	Potential Reduction (%)	Adjusted Operating Time (%)	Potential Reduction (%)	Adjusted Operating Time (%)
Setup and gauging	22	30	15.4	65	7.7
Load/unload, noncutting	12	40	7.2	60	4.2
Tool change	22	5	20.9	15	18.7
Equipment failure	7	0	7	0	7
Idle time	12	0	12	25	9
Direct processing of materials	25	0	25	0	25
Total[b]	100		87.5		64.6
Potential throughput index	1.0		1.14		1.55

[a]Breakdown of theoretical capacity from Mayer and Lee, 1980a.
[b]Total scheduled production time.

were consolidated and controlled by sensor-based computer systems. For example, sensors monitoring machine performance would eliminate unnecessary adjustments and would speed up diagnoses of machine failures.[3] If stand-alone machines were replaced by a flexible manufacturing system, and part design and processing were rationalized using group technology principles (Ham, 1984; Houtzeel, 1985), there would be less material handling and the scheduling of parts and tools would be simplified. Even a substantial fraction of the equipment-related losses could be eliminated in a fully integrated flexible manufacturing system, since the whole system need not be stopped if one station malfunctions. Robots or programmable pallets under the control of a central scheduling computer could reroute parts to other work stations.

Without increasing the time normally planned for operation and without making extensive additions of other forms of automation, these calculations indicate that the installation of robots would result in a 10 percent increase in throughput in high-volume machining operations, and nearly a 15 percent increase in throughput in medium- and low-volume production.[4] If robots were used in

3. In the next few years, time lost to equipment failures could conceivably increase as systems become more automated and more complex. However, improvements in machine reliability and in sensor-based diagnostic systems could improve machine and system reliability and reduce equipment failures over the next two decades.

4. The 15 percent increase in medium- and low-volume production derived from the breakdowns of theoretical capacity coincides with the 10–20 percent increase derived from the interviews with two experienced robot users. It appears that my "informed" judgments are consistent with actual experience.

Table 4-9: Potential Increases in Throughput from Working More Efficiently During Scheduled Production Time: Low-Volume Manufacturing

Function	Operating Time[a] (%)	Robots Only		Robots + CAM	
		Potential Reduction (%)	Adjusted Operating Time (%)	Potential Reduction (%)	Adjusted Operating Time (%)
Setup, loading and gauging	55	25	41.3	50	27.5
Idle time	9	0	9	50	4.5
Cutting conditions	9	0	9	25	6.8
Direct processing of materials	27	0	27	0	27
Total[b]	100		86.3		65.8
Potential throughput index	1.0		1.16		1.52

[a]Breakdown of theoretical capacity: from Mayer and Lee, 1980a.
[b]Total scheduled production time.

conjunction with other forms of factory automation systems while maintaining the same number of days normally planned for operation, throughput might be increased by nearly 50 percent in medium- and low-volume production, and possibly by 40 percent in high-volume production. These estimates suggest that retrofitting robots into existing production lines will bring about some improvements, such as improving the utilization of a single machine or work station, but will not dramatically improve overall factory performance. I suggest here that substantial impacts on performance at the factory level require the integration of robots and other forms of factory automation into coordinated manufacturing systems. This is a view commonly expressed by industry specialists (Yoshikawa, Rathmill and Hatvany [1981], Bylinski [1983], National Research Council [1984]).

The potential combined effects of fully utilizing unscheduled production time and of working more efficiently during scheduled operations are shown in Table 4-10. The current level of output, given the amount of time typically scheduled for production and the amount of time typically spent processing materials, is designated as the base case. The available hour index gives the increase in output that would result from fully utilizing time not normally scheduled for production. The throughput index is the increase in output that would result from working more efficiently. The magnitude of the throughput index depends on whether robots are used alone or integrated with other types of CAM technologies. The total output index is the product of the available hour index and the throughput index.

If it were possible to operate around the clock in the robots-only case, the potential increase in output would be nearly 50 percent for high-volume operations, over 150 percent for low-volume, two-shift operations, nearly 250 per-

Table 4-10: Summary of Potential Increases in Output

Type of Plant	Potential Capacity Increases		
	Base Case	Robots Only	Robots + CAM
High-volume			
Available hour index	1.00	1.31	1.31
Throughput index	1.00	1.11	1.39
Output index	1.00	1.45	1.82
Increase in output (%)		45	82
Low-volume two-shift			
Available hour index	1.00	2.17	2.17
Throughput index	1.00	1.16	1.52
Output index	1.00	2.52	3.30
Increase in output (%)		152	230
Medium-volume			
Available hour index	1.00	2.98	2.98
Throughput index	1.00	1.14	1.55
Output index	1.00	3.40	4.62
Increase in output (%)		240	362
Low-volume, one-shift			
Available hour index	1.00	4.35	4.35
Throughput index	1.00	1.16	1.52
Output index	1.00	5.05	6.61
Increase in output (%)		405	561

Sources: Available hour indices derived from Table 4-6. Throughput indices derived from Tables 4-7, 4-8 and 4-9.

cent for medium-volume operations, and over 400 percent for low-volume factories operating on a single shift. To utilize all of the time normally not scheduled for production (plant shutdowns, holidays, Sundays) the plant would sometimes have to operate with skeleton crews, or even operate unmanned during some periods under the control of computer systems. If it is assumed that production time is increased to its theoretical upper limit, there is a rationale for arguing that only the case of robots used in conjunction with other CAM technologies makes sense. For the case of robots with CAM, high-volume operations show a potential output increase of 80 percent. Producers already on a double shift and medium-volume manufacturers show a potential output increase of 230 percent and 360 percent, respectively. For low-volume producers operating on normal single shifts, the potential increase is 560 percent. In medium- and low-volume manufacturing, potential increases in output are almost all the result of increasing planned production time. In high-volume machining operations, the contribution of an increased throughput rate is somewhat greater than the contribution of increased hours planned for production.

In summary, these calculations suggest that if a plant currently organized to intermittently produce small and medium-sized batches of specialized products were reorganized to produce in a more continuous fashion around the clock, the level of output in the reorganized facility would greatly exceed that of the conventional facility. The data presented here indicate that the level of output could potentially be increased by 150–550 percent.

4.4.1. Utilization of Metalcutting and Metalforming Tools

Estimates of the average utilization of all metalcutting and metalforming machines in SIC 34–37 are shown in Table 4.11. The definition of machine utilization was already introduced and discussed in Chapter 3. The estimates for all cutting and forming machines averaged together range from a low of 11.2 percent in SIC 36 to a high of 16.7 percent in SIC 37. These estimates suggest that, on average, cutting and forming machines are used only one-sixth to one-ninth of the time they are theoretically available. As mentioned in the discussion of utilization in Chapter 3, comparing total operator hours with total available machine hours for estimated levels of machine utilization implicitly assumes that each machine is equally important in the production process. In general, this is not the case. Older (and often fully depreciated) machines are typically not discarded, even though they are not used on a regular basis. One-third of all metalcutting and metalforming machines in factories are over 20 years old according to the 12th *American Machinist* Inventory. Older machines are often kept in reserve for surge capacity, or are permanently set up to perform unusual jobs that occur only occasionally.

If total machine hours available were reduced by one third (to adjust for the fact that many of the older machines are intentionally left idle), it would appear that cutting and forming machines are used only from one-fourth to one-sixth of the time they are available, suggesting that output could be increased by a factor

Table 4-11: Estimated Percentage of Average Utilization of Metalcutting and Metalforming Machines, 1977

Major Group	Metalcutting Tools, Manual	Metalcutting Tools, NC	Metalforming Tools	Average, Cutting + Forming
34	12.4	18.4	15.8	13.9
35	12.7	20.9	7.4	12.3
36	9.9	29.3	13.0	11.2
37	16.5	14.8	17.5	16.7
Average	12.9	20.0	13.6	13.1

Sources: number of machines: *American Machinist,* 1978; number of operators: Bureau of Labor Statistics, 1980; hours worked by production workers: *1977 Census of Manufactures* (Bureau of the Census, 1981c, d).
Assumptions:
Machine utilization is defined here as follows:

$$\text{utilization} = \frac{\text{total operator hours available}}{\text{total machine hours available}}$$

where
total operator hours available =
  number of operators × average hours worked per operator per year
average hours worked per operator per year =
  total hours worked by production workers/total production workers
total machine hours available = number of machines × 8760 hours per year
8760 hours per year = 24 hours per day × 365 days per year

of four to six. Table 4-11 shows that the utilization rates of numerically controlled (NC) metalcutting machines in SIC 34, 35, and 36 are substantially higher than the overall rate of machine utilization in these industry groups. This suggests that in these groups, the newer, more productive machines are used to perform a disproportionate share of the work. This is consistent with the hypothesis that many older machines are not used for production purposes. However, even in SIC 34, 35, and 36, the NC machines are in use only from one-third to one-fifth of the total time they are available.

The conclusion derived from a comparison of operator hours with machine hours is the same as the conclusion drawn from the Machine Tool Task Force breakdown of theoretical capacity. *The utilization of machine tools could be increased severalfold.* The conclusion derived from both data sources concurs with the results of a worldwide survey of manufacturing engineers and scientists reported by Merchant (1983). Using present-day standards as a basis, respondents were asked to estimate the ultimate percentage of change that computer-based automation, optimization, and integration in the metalworking industries will achieve in several areas of performance (productivity, quality, lead times, utilization, and inventory). The median response for the projected increase in levels of machine utilization was 340 percent, with answers ranging from 20–1500 percent.

Figure 4-1 shows a graph of pounds of metal processed per establishment versus the utilization of all metalcutting and metalforming tools. The increasing trend in the data supports the assertion that if levels of machine utilization were increased, levels of output could be increased substantially. In fact, the regression results indicate that there is more than a 1 percent increase in output per establishment for every 1 percent increase in the utilization of cutting and forming machines.

4.5. The Feasibility of Increasing Output Levels Severalfold in Batch Production Facilities—The Example of FMSs in Machining

Despite the analysis of the theoretical potential for increasing output by fully utilizing available production time, one wonders if it is really possible to have a plant operating around the clock, sometimes unmanned, continuously producing a mix of different types of products. Recent experience with the use of flexible manufacturing systems (FMSs) in machining operations indicates that this *is* possible to achieve, at least within a subset of manufacturing operations. For example, Jaikumar (1986) reports that of 60 FMSs he studied in Japan, 18 ran untended during the night shift. Warnecke and Steinhilper (1985: 5) define an FMS as follows:

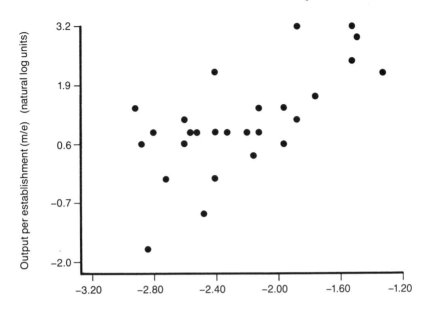

Utilization of metalcutting and metalforming machines (u)
(natural log units)

THE REGRESSION EQUATION IS
ln (m/e) = 4.79 + 1.66 ln(u)

Variable	Coefficient	St. dev. of coef.	T-ratio = coef./S.D.
constant	4.7937	0.8157	5.88
ln(u)	1.6601	0.3574	4.65

The st. dev. of Y about regression line is S = 0.8411 with (27 − 2) = 25 degrees of freedom
R^2 = 46.3 percent
R^2 = 44.2 percent, adjusted for d.f.

Figure 4-1: Output per Establishment versus Utilization of All Metalcutting and Metalforming Machines for Three-Digit SIC Industry Groups, 1977

A *flexible manufacturing system* contains several automated machine tools of the universal or special type, and/or flexible manufacturing cells and, if necessary, further manual or automated work stations. These are interlinked by an automated workpiece-flow system in a way that enables the simultaneous machining of different work pieces which pass through the system along different routes. Thus automated, multi-step and multi-product manufacturing is possible in a flexible manufacturing system. Setting times of system components are designed such that the undisturbed operation of the other components is guaranteed during a setup process. The actual operation of flexible manufacturing systems shows an exploding growth—there are now approximately 250 flexible manufacturing systems installed worldwide. It is expected that the number of installations will double every two years until the end of the decade.

Table 4-12: Comparison between the Performance of the FMS Operation at Toshiba Tungaloy Company and a Conventional Machining Operation

Measure of Resource Usage or Performance	Conventional Machining Operation[a]	FMS[b]
Labor requirements (3 shifts)	70 persons	18 persons
Number of machine tools	50 sets	6 sets
Number of machining processes	15	8
Machine tool utilization	20%	73%
Product yield	95%	98%
Average production lead time	18.6 days	4.2 days
Shop floor space	1480 m^2	350 m^2

Products manufactured: carbide tools and alloy tips. Facility: Kawasaki, Japan, plant (information from Merchant, 1985).
Source: Sekiguchi, 1985: 124.
[a]Values for the conventional machining operation are estimates made by company engineers based on past experience with conventional stand-alone NC and manually controlled tools.
[b]Values for the FMS are actual measurements.

Comparisons of resource requirements and performance of batch-production operations using FMS and conventional technologies are shown in Tables 4-12 and 4-13. Table 4-12 describes the FMS implemented by the Toshiba Tungaloy Company, Ltd., a Japanese manufacturer of carbide tools and alloy tips used in metalcutting. Sekiguchi (1985) reports that concept planning for the system began in 1977 and that regular three-shift operations began in mid 1981. Merchant (1985) reports that the FMS can handle some 4000 different part types, that lot sizes range from 2 to 20 pieces, and that part of the system runs unattended during the third shift. Table 4-12 also compares system resource and efficiency measures of the FMS with those of a conventional system with approximately equal production capabilities. Sekiguchi reports that the data for the FMS are actual measurements and that the data for the conventional system are estimates based on operating experience. Merchant reports that the data for the conventional system were obtained from a comparison made between the performance of the FMS and a performance obtained in another Tungaloy factory producing a similar range of cutting tools with a mix of conventional and NC machines used in stand-alone fashion.

To achieve the specified level of output and product mix produced by the FMS described in Table 4-12, a conventional system would require nearly four times as many people, nearly twice the number of machining processes, and many more machine tools. The machines in this FMS tend to be large, expensive, multifunctional machining centers which incorporate the functionality of several types of stand-alone machines into one unit. Essentially, different types of machining operations have been consolidated into large machining centers. The significant aspect of this type of consolidation is that the need to set up, as well as to load and unload machines has been reduced, and therefore the time

Table 4-13: Comparison between the Performance of the FMS at Yamazaki Machinery Works and a Conventional Machining Operation

Measure of Resource Usage or Performance	Conventional Machining Operation[a]	FMS[b]
Labor requirements (3 shifts)	106 persons	12 persons
Shift 1	40	6
Shift 2	40	6
Shift 3	26	0
Number of machine tools	36 sets	18 sets
Number of machining processes	10	4
Number of cutting tools used	base	1/6 of base
Time to machine a single large part	33 hours	14 hours
Average production lead time	90 days	4 days
Level of in-process inventory	base	1/5 of base

Products manufactured: parts for NC machine tools and robots. Facility: Oguchi, Japan, plant (information from Hartley, 1985a).
Source: Sekiguchi, 1985: 125–131.
  [a]Values for the conventional machining operation are estimates made by company engineers based on past experience with conventional stand-alone NC and manually controlled tools.
  [b]Values for the FMS are actual measurements.

required to carry out machining operations has also been reduced substantially. Levels of machine utilization are nearly four times greater in the FMS than they would be in a conventional system. The average total processing time to go from raw material stock to a finished product (the production lead time) has been reduced from 18.6 to 4.2 days. Sekiguchi (1985) reports that this decrease was the result of reducing setup times and the number of operations.

Performance data for the FMS used by Yamazaki Machinery Works, Ltd., a Japanese producer of NC machine tools, are shown in Table 4-13. Sekiguchi reports that concept planning for the system began in 1978, and by mid-1983 the system was in full operation at 60 percent of its potential capacity. Sekiguchi (1985: 130) provides a description of the system's operation.

Six employees are needed for day and night shifts—one in the tool presetting room, one in the computer room, and two in each loading station. Therefore, only twelve employees work during the two shifts.

The number of machining processes decreased from ten to four, mainly because of the reduction in the number of setup operations created by using machining centers and changing product design. Moreover, the reduction of setup times reduced total average operation time by almost half. For example, total operation time to produce a large-size part dropped from thirty-three hours to fourteen.

As in the previous example, the new automated system can produce a given level of output with a fraction of the labor input and in a fraction of the time that would be required in a conventionally organized batch-production facility. According to Bylinski (1983), investment costs for the system totaled $20 million. It is not known whether this amount includes the cost of the 100,000 engineering man-hours that Sekiguchi reports were needed for planning and designing the system. Bylinski reports that the Yamazaki management expects that over time the new FMS will be 15 times more profitable than a conventional plant its size. The company estimates that after five years of operation, the plant will produce after-tax profits of $12 million compared with an estimated $800,-000 for a conventional batch-production facility. Kaplan (1986) describes this same facility, although his description of the performance comparisons does not match those given by Sekiguchi in Table 4-13. Kaplan reports that after two years, total savings came to only $6.9 million, $3.9 million of which had flowed from a one-time cut in inventory. Even if the system were to continue to produce annual labor savings of $1.5 million for 20 years, the project's return would be less than 10 percent per year.

Hartley (1985a) describes Yamazaki's more recent FMS machining operation in Minokamo, Japan, which uses 43 machines and 17 robots. A summary of the part variety and potential output for the major part families produced at the facility is given below:

Part Families	Number of Different Part Types Produced	Potential Output (units/month)
Flanges, spindles, and other small parts	435	9000
Gear boxes	85	1200
Large frames	40	240

Hartley (1985a: 278) provides the following description of the system's operation:

In 1983, the plant was operated 15 hours a day with 240 people, and Yamazaki claimed that this was solely due to the slack market. Only 39 people are needed to run the FMS, with three men/shift on the spindle line, three men/shift on the flange line, five on the box line, and four on the frame line. Three people are needed to load work pieces on to the Automated Guided Vehicles (AGVs) outside the shop, and there is an operator in the computer room, and another in the tool room. The system would be virtually unmanned on the third shift, and clearly has the potential for an unmanned second shift. About 80 people work on assembly.

According to Hartley, Yamazaki management claims that the facility has a potential output of three shifts and that it is comparable to a conventionally organized batch-production facility with 120 lathes and machining centers. Excluding the land, Yamazaki made an investment of 11 billion yen for the plant.

Perhaps the most extensive efforts toward building plants that can operate around the clock while producing a mix of different products are the recent facilities built by Fanuc, the Japanese manufacturer of NC machine tools and robots. A description of the first major FMS built by Fanuc, in Fuji, Japan, is given by Yoshikawa, Rathmill, and Hatvany (1981: 11–12).

> Recently, there has been a remarkable trend toward having systems unmanned during the night shifts. Fanuc has constructed a new factory in Fuji that does this. This plant has 29 cell-like work stations. Seven of them are equipped with robots; 22 are equipped with automatic pallet changers with pallet pools. These stations are connected by unmanned vehicles guided by electromagnetic or optical methods. The plant has two automatic warehouses, one for materials and another for finished parts and subassemblies. The vehicles transport the materials from the warehouse to the unmanned machining stations. Robots or automatic pallet changers load materials onto the stations from the vehicles. Finished parts are transferred again automatically by the vehicles to the second warehouse.
>   The plant has an assembly floor, where workers work during the day. Transportation between the parts warehouse and this floor is also by unmanned vehicles. In the daytime, 19 workers are working around machining stations, mainly for palletizing, and 63 workers are in the assembly section. Thus, 82 workers are in the plant during the day, but at night there is only one. The assembly floor is closed, and the machining floor is operated without any workers. Every station is equipped with a monitoring device with a TV camera, and one person sits in the control room to monitor all the working stations. He observes the working status of stations through the camera, without touring the factory. The monitoring device also records the spindle motor current, calculating the cutting force and time to judge cutting conditions.
>   Many Japanese machine tool manufacturers have recently completed commercial systems with monitoring devices that can operate nearly unmanned night shifts.

According to Bylinski (1983), the investment is estimated at $32 million. Bylinski reports that Fanuc management claims that a conventionally organized batch-production facility would require 10 times the number of workers and capital investment to provide the same output and product mix as the FMS. The plant is five times more productive than its conventional counterpart would be.

Hartley (1985b) describes a more recent Fanuc facility in Japan which produces servo motors. It was build at a cost that is "a little less than was invested in the first Fuji factory." Hartley provides the following comments on the plant's operation and performance:

> The new factory is a compact two-story building, with a machine shop on the ground floor, and assembly on the first floor. It is used to produce spindle motors, ac and dc servo motors. Capacity is put at 10,000 motors of 40 kinds per month, but output is currently much less. Nevertheless, 900 different parts are machined in lot sizes from 20 to 1000. . . . In all, 21 men work on one shift in the machine shop, and their main job is to maintain the machines and robots. The machines operate unmanned for the other two shifts, with just one man in the control room. . . . Although the machine shop operates for 24 hours/day, the assembly shop is operated for only 8 hours/day. The reason is that

some of the assembly is done manually. The level of automated assembly is put at 65%.
. . . In the complete plant, there are 60 employees, including 21 in the machine shop—on
one shift—and 19 direct workers in the assembly shop. . . . All the Japanese electrical
companies are now trying to increased productivity, and in theory, the new plant
increased productivity three times. Fanuc used to have 32 robots and 108 men to produce
6000 motors [per month]; now 60 men and 101 robots produce 10,000 motors. . . . The
manufacturing cost of the servo motor will be cut by 30% including depreciation. (pp.
282–286)

   Flexible manufacturing systems for machining operations have also been in
commercial use in the United States for about 20 years.[5] Examples of systems
that have been installed in the United States are given in Table 4-14. Additional
information on investment in hardware and reported savings in resources for
several systems is given in Tables 4-15 and 4-16. Investment in system hardware
runs into the multimillion-dollar range. Users report large savings in operating
costs, including reductions in labor cost and in costs associated with in-process
inventories and with floor-space requirements. Another performance comparison between FMSs and conventional machining operations made by General Electric is shown in Table 4-16. The hardware cost of the system was about
$16 million. The FMS has a third as many machine tools as the system it
replaced. Twenty-nine manually operated machines were replaced by nine automated machining centers. As a result, floor-space requirements were reduced
by 25 percent, and the typical number of times a part had to be loaded onto a
separate machine was cut in half. The total number of people required to support
the machining activity over two shifts (material handlers, operators, maintenance workers, schedulers, and supervisors) was reduced from 86 to 16. GE
management reports a 240 percent increase in total factor productivity as a
result of installing the new system, although they do not specify over what time
period this increase has been (or will be) realized.
   Several points from these descriptions of FMS applications are worth under-
scoring. The first is, these examples show that with the most advanced types of
automation currently available, aspects of the manufacturing process for the
production of batches of specialized products can indeed be *fully* automated. In
the Yamazaki and Fanuc plants in Japan, production workers maintain, super-
vise, and program the machines. They are not involved in the direct processing
of materials. Workers on the second shift perform setup and preparatory functions that are required for the plant to run unmanned during the third shift. It
should be emphasized that in all the applications described only the machining

5. Jaikumar (1986: 69) notes: "Compared with Japanese systems, those in U.S. plants produce an
order-of-magnitude less variety of parts. Furthermore, they cannot run unattended for a whole shift,
are not integrated with the rest of their factories, and are less reliable. Even the good ones form, at
best, a small oasis in a desert of mediocrity."

Table 4-14: Partial List of FMSs for Machining in Use in the United States

User	Vendor	Date Installed	Total Number of Parts Machined per Year	Number of Different Part Types Machined	Type of Part
Sundstrand Aviation	White-Sundstrand	1967	na	na	aluminum pump parts
Ingersoll-Rand	White-Sundstrand	1970	20,000	14	hoist and motor cases
Rockwell	K&T	1972	25,000	45	automotive axle carriers
Allis-Chalmers	K&T	1970–73	23,000	8	agricultural equipment
Caterpillar	White-Sundstrand	1973	6,600	6	crank case housings, covers
AVCO-Williamsport	K&T	1975–78	24,000	9	aircraft engines
AVCO-Lycoming	K&T	1979	15,000	10	turbine engines
Caterpillar	Giddings and Lewis	1980	na	na	construction equipment
John Deere	K&T	1981	5,000	8	farm equipment
International Harvester	Giddings and Lewis	1981	na	na	na
General Electric	Giddings and Lewis	1981	5,600	7	motor housings
AVCO	Giddings and Lewis	1982	na	na	na
Detroit Diesel Allison	White-Sundstrand	1983	na	40	transmission housings

Source: Draper Lab, 1983.
na = information not available.

Table 4-15: Hardware Investment and Reported Savings in Operating Costs for Selected Kearney & Trecker Machining FMSs

User	Cost (in millions of dollars)[a]	Total Number of Parts Machined Per Year	Number of Different Part Types Machined	Comparisons with Old System
Rockwell (truck axles)	$5.6	24,000	45	¼ floor space; setup costs virtually eliminated
AVCO-Williamsport (aircraft engines)	$8.4	24,000	9	⅓ floor space; ¼ labor; ½ number of part-holding devices
John Deere (tractor components)	$18	50,000	8–12	cost estimate of: FMS: $18 million; dedicated transfer line: $28 million
Mack Truck (truck components)	$5	65,000	5	cost estimate of FMS about the same as estimate for dedicated transfer line, with comparable cycle time, less flexibility
Caterpiller[b] (construction equipment)	$5	8,000	8	total transit time through system: old system: 8.5 hours new system: 0.3 hours

Source: Klahorst, 1983a.
[a]1982 value.
[b]This system is a semiflexible transfer line. All other systems listed here are FMSs.

Table 4-16: Performance Comparison between the FMS and Conventional Machining Operation at General Electric

	Conventional Machining Operation	FMS
Number of machines	29	9
Total production worker requirements for two shifts (operators, supervisors, and maintenance)	86	16
Typical number of machine loadings required to complete one part	10 or 11	4 or 5
Maximum annual output for family of seven parts	4100	5600
Average production lead time	16 days	16 hours
Cost	base	$16 million
Total factor productivity change reported by GE management	base	+240%

Product manufactured: motor frame housings for locomotives. Facility: Erie, Pennsylvania, plant.
Sources: Miller, 1983; Manufacturing Engineering, 1983: 66–67; and Bylinski, 1983.

of parts is fully automated, and that the machined parts are assembled in more conventional ways requiring human workers. However, in the Fanuc plant at Minokamo, even a large fraction of assembly operations are automated. This indicates that the range of manufacturing operations that can be automated with the so-called flexible type of system is gradually expanding.

Second, in each of these systems, the potential output is several times that of its conventional counterpart. For example, in the comparisons shown in Tables 4-12, 4-13, and 4-16, a given level of output and product mix is achieved with the FMS with substantially fewer people than would be required if a conventionally organized batch-production facility were used. Also, using a small number of multifunctional machining centers as opposed to a large number of conventional machine tools substantially reduces time spent for setups and for loading and unloading. For these and other reasons, utilization rates are higher. The data for the Tungaloy system show that utilization levels in the FMS are nearly four times those typically achieved with conventional systems. The National Research Council (1984) reports that in five U.S. companies which have been leaders in applying computer-integrated manufacturing the productivity of capital equipment (the time it actually spends in operation) has increased from two to three times. Companies using new types of automated systems claim that they are much more productive than the operations they replaced. The GE management claims that its system is two and one-half times more productive than the system it replaced, and the Fanuc management claims that its new batch-production plants are three to five times more productive than their conventional counterparts. *A critical point is that the magnitude of the potential for increasing output levels in batch-production facilities inferred from the information on the FMS applications is consistent with the estimates derived in the preceding Section 4.4.*

The third notable point is that robotic manipulators are only a very small part of the total automation used in these systems. Insensate (Level I) robotic manipulators are sometimes used in large FMSs to load and unload parts, as indicated in the description of the Fanuc plant in Fuji. A robot is used in the GE system to move cutting tools into and out of a machine tool. Seventeen robots are used in the Yamazaki plant in Minokamo to load work pieces onto pallets, and also to remove metal chips and shavings from parts. In the new Fanuc plant, most of the NC machine tools are loaded and unloaded by robots. In general, however, robotic manipulators do not play a major role in these types of systems, primarily because the basic transformation task, the machining, is already automated by the machine tool. Most of the other tasks are related to material handling, and there are often alternative forms of automation to move parts and tools. For example, automated guided vehicles are often used to transport parts from the loading station to the machine tool. There are also many ways of moving a tool from a storage rack to the tool compartment in the machine tool.[6] The cost of robots or robotlike manipulators constitutes only a negligible fraction of the multimillion-dollar investment required to implement these systems. This suggests that when analyzing the case of a fully utilized plant running around the clock one should more appropriately address the potential impact of flexible automation *systems* on cost and labor requirements, as opposed to the impact of robotic manipulators. In operations such as assembly, where the robot performs the basic transformation task (parts mating), robots would constitute a larger fraction of total system cost. Nonetheless, it would still be the case that the impacts on cost and labor would be due to a system of automation rather than the use of robotic manipulators.

The fourth point is that very large capital investments are required to design and install such systems. The hardware costs of the Kearney & Trecker systems listed in Table 4-15 range from $5 million to $18 million. This sum includes the cost of the machine tools, material handling equipment, control systems, tools and fixtures, testing, training, manuals, and other support services. This sum does not include the costs of planning, engineering, and installation. There are no data available on these support costs. However, Kearney & Trecker (1982) has found that it typically takes three years of planning from the time a customer decides to buy an FMS until the system is installed. Considering that almost every functional area of manufacturing is affected by the use of an FMS and must be involved in planning to some extent, it is inevitable that the costs of planning and engineering are considerable. The costs of the Japanese FMSs are reported to be in the $20–$30 million range. These cost figures must be interpreted cautiously, since it is not known whether they include all planning, design, and software costs, or only the cost of the plant and equipment. As noted earlier, Sekiguchi (1985) reports that the Yamazaki FMS described in Table 4-13 required over 100,000 hours of engineering effort for the design of machines, equipment, and electrical and electronic systems, and for the integration of the total system design. Klotz (1984) reports that the plant built by Yamazaki's U.S. affiliate, Mazak, in Florence, Kentucky, in the early 1980s cost over $200 million, exclusive of planning and engineering costs. The plant, which builds machine tools, has five FMSs, with a total of 60 computer NC machine tools. Klotz estimates the design required hundreds of thousands of engineering hours. Clearly, when considering investment requirements of this magnitude, capital costs cannot be ignored, as they were in the examples discussed in Sections 4.1 and 4.2.

6. Interestingly enough, robots are used more extensively in manufacturing cells, which are smaller than the large-scale FMSs. If the machines in the cell are properly arranged, it is typically less expensive to use one or several robots to move materials from one machine to another than to use other types of material-handling equipment.

The fifth point is that while large capital investments are required to build fully integrated FMSs, there are indications that in some cases it would cost even more to build and operate a plant of comparable capabilities if conventional types of machinery were used. If conventional types of stand-alone, NC, general purpose machines were used, all aspects of materials processing would not be fully automated, and setups would be required frequently. Throughput rates would not be as high as in FMSs, so more workers and machines would be required to achieve a given capacity. If conventional types of fully automated, high-volume transfer lines were used, each line could produce only one type of product, since the control processes at each work station are specifically designed for one type of operation. More than one type of conventional transfer line might be required for each flexibly automated line. Table 4-15 shows the two cases where the price quote for the FMS was equal to or less than the price quote for a special-purpose transfer line. This indicates that there are some situations where the use of FMSs over conventional alternatives can actually reduce the investment required to automate while providing the advantages of flexibility. Fanuc management claims that in the Fuji plant several times more capital investment would have been required to achieve the comparable level of output. However, for the Yamazaki system in Table 4-13, the investment was apparently more as a result of using the new type of automation. Sekiguchi (1985) reports that this FMS cost about 4 billion yen, compared with an estimated 3 billion yen for a production facility with almost the same production capacity, thus indicating that the flexibly automated system cost one-third more than the conventional alternative would have.

The last point is that the descriptions of FMS operations indicate only the reductions in production worker requirements and give no information on the effects on other positions in the factory, such as engineering and technician jobs. Information on changes in total labor requirements is required before making a conclusive statement on the decrease in total labor costs.

4.6. A Framework for Estimating Reductions in Unit Cost Resulting from a Severalfold Increase in Output in a Batch-Production Facility

The extent to which a severalfold increase in output would decrease unit cost in a fully utilized batch-production plant is not known. Very few empirical data are available to make such a comparison, since there are only a handful of flexibly automated plants throughout the world. The information that has been published on the relative capabilities of conventional and flexibly automated plants is too sparse to facilitate a detailed analysis. Given that the unit cost in the proposed high-volume batch-production plant cannot be directly observed, it must be inferred or approximated through some indirect means. The framework used here for estimating the potential reduction in unit cost is to assume that a flexibly automated plant producing specialized products in small and medium-sized batches would have some of the characteristics of conventional plants producing more standardized products in larger volumes. Using the data set discussed in Chapter 3, the measures of value added per pound of metal processed and of pounds of metal processed per number of establishments are used to estimate a regression relationship between cost per unit and units of output. This estimate of unit price elasticity with respect to output is used to calculate the percentage of reduction in unit cost which would result if the output in a metalworking plant were increased by factors of 1.5–10.

One feature of this analysis is that the measure of value added for each industry incorporates both labor and capital costs. The scale elasticity derived from the regression relationship implicitly incorporates the effect of changes in both unit capital cost and unit labor cost across metalworking industries as output is increased, even though the incremental cost of purchasing additional robotic capital is not treated explicitly. Since the data set is for the year 1977, and very few, if any, of the establishments were using FMSs then, the capital cost increases are based on the use of conventional types of capital equipment. This is a point of concern if the cost of the quantity of flexible automation required to increase output would be less than the cost of using the quantity required of conventional types of equipment, as is suggested by the discussion of the Japanese systems. No attempt is made to correct for this possibility, although the implications of a possible difference in the cost of flexible and conventional automation is discussed later on, when the estimates of the output elasticity are reviewed.

4.6.1. A SUMMARY OF THE ESTIMATED ELASTICITY OF UNIT COST WITH RESPECT TO OUTPUT

A more complete discussion of the expanded regression model used to estimate the elasticity of unit cost with respect to output is given in Appendix E. Only the results are presented here, along with an overview of the model. Justification for the use of additional variables to account for interindustry differences in the nature of the processing requirements is given in Appendix E.

The explanatory variables used in the expanded multiple regression model are summarized in Table 4-17. The proposed effect on unit cost with respect to each variable is also shown by indicating whether the sign of the estimated elasticity of unit cost should be positive or negative. The sign is determined by assuming that there is a 1 percent increase in the particular explanatory variable. The sign should be positive if a 1 percent increase in the explanatory variable corresponds with an increase in unit cost and negative if the increase corresponds with a decrease in unit cost. Unit cost should decrease as the level

Table 4-17: Summary of Explanatory Variables in Multiple Regression Model

Factor	Variable	Notation	Effect on Unit Cost (sign of coefficient)	
Level of output	$\dfrac{\text{pounds of metal processed}}{\text{number of establishments}}$	m/e	$\dfrac{d \ln(u)}{d \ln(m/e)}$	< 0
Complexity of metal-shaping activities	$\dfrac{\text{dollars of basic metal}}{\text{pounds of basic metal}}$	bmci	$\dfrac{d \ln(u)}{d \ln(bmci)}$	> 0
Degree of assembly	$\dfrac{\text{dollars of purchased metal}}{\text{dollars of basic metal}}$	pmbm	$\dfrac{d \ln(u)}{d \ln(pmbm)}$	> 0
"All included" hourly wage	$\dfrac{\text{production worker wages + benefits}}{\text{production worker hours}}$	w	$\dfrac{d \ln(u)}{d \ln(w)}$	> 0
Material coverage	$\dfrac{\text{dollars of metals}}{\text{dollars of total materials}}$	c	$\dfrac{d \ln(u)}{d \ln(c)}$	< 0

THE REGRESSION EQUATION IS

$$E[\ln(u)] = b_0 + b_1\ln(m/e) + b_2\ln(bmci) + b_3\ln(pmbm) + b_4\ln(w) + b_5\ln(c)$$

WHERE

$E[\ln(u)]$ = expected value of $\ln(u)$

u = unit cost = value added/pounds of metal processed

b_i = ordinary least squares estimate of β_i

of output increases, so the sign of the coefficient for pounds of metal processed per number of establishments should be negative. The average unit cost of the basic metals purchased is called the basic metal cost index. An increase in the index means that more expensive metals are used, which is taken as an indication of increased difficulty in the shaping operations. Since more difficult operations require more capital and/or labor inputs, unit cost is assumed to increase and the proposed sign of this elasticity is positive. There are two reasons for believing that a higher ratio of processed metal cost to basic metal cost indicates a more complex process. First, the higher the ratio of processed metal inputs to basic metal inputs, the greater the diversity of material inputs used in an industry. There is some tendency for the ratio of salary costs to production worker cost to increase across industries as the ratio of processed metal inputs to basic metal inputs grows larger. This provides some evidence for arguing that it takes more organizational control and supervision to coordinate production when there is a larger proportion of processed metal inputs. Second, the inspection of assembled products requires more than just the verification of dimensions. Since subcomponents must be properly integrated with one another, testing is required to verify that the final product correctly performs its designated functions. With electronics equipment and computers, this can be a fairly extensive and complicated process. Given these assumptions, unit cost should increase

Table 4-18: Bonferroni Joint Confidence Intervals for Output and Complexity Parameters, Total Data Set SIC 34–37, n = 101

Parameter for		
ln(m/e)	ln(bmci)	ln(1 + pmbm)
$-0.3867 < \beta_1 < -0.2029$	$0.7669 < \beta_2 < 1.1985$	$0.3718 < \beta_3 < 0.6044$
$b_1 = -0.2948$	$b_2 = 0.9847$	$b_3 = 0.4881$

Family Confidence Coefficient = (1.0 – 0.05) = 0.95

See Table 4-17 for definitions of variables and full regression model.
Intervals are derived from regression results in Appendix E.

with an increase in the ratio of dollars of processed metals to dollars of basic metals, and the elasticity of unit cost with respect to this variable should be positive.

Bonferroni joint confidence intervals are specified for the following parameters in Table 4-18: pounds of metal per number of establishments (β_1), the basic metal cost index (β_2), and the proportion of assembly to metal-shaping activities (β_3). Only these three parameters are examined in detail, because the issue of concern here is the relative influence of the level of output and of the nature of the processing requirements on unit cost.

In the expanded regression model, the expected value of the unit cost elasticity with respect to output, b_1, is –0.295. In Table 4-19, this estimate of the unit cost elasticity is compared with other estimates which have been referred to in Chapter 3. The output elasticity estimated from the cross sectional data (without including the other explanatory variables)—Case B in Table 4-19, falls right in the middle of the Lamyai, Rhee, and Westphal (1978) estimates, which are designed to encompass the lowest-cost and highest-cost estimates of manufac-

Table 4-19: Comparison of Estimated Elasticity of Unit Cost with Respect to Output for Metalworking

Source	Expected Value of Estimate	Confidence Interval for Output Elasticity $(1 - \alpha = 0.95)$
Case A: Miller[a] Unit cost vs. scale *and* processing complexity	–0.295	$-0.387 < \beta_1 < -0.203$
Case B: Miller[b] Unit cost vs. scale only	–0.436	$-0.562 < \beta_1 < -0.309$
Case C: Lamyai, Rhee, Westphal[c] Full-sharing case	–0.362	$-0.400 < \beta_1 < -0.324$
Case D: Lamyai, Rhee, Westphal[c] No-sharing case	–0.524	$-0.564 < \beta < -0.484$
Case E: Cook[d]	–0.392	$-0.450 < \beta_1 < -0.333$

[a]See Appendix E. Based on an analysis of a cross section of industries.

[b]See Chapter 3. Based on an analysis of a cross section of industries.

[c]See Chapter 3. Based on analysis of one type of product.

[d]See Chapter 3. Based on analysis of one type of product. An output elasticity is not actually estimated by Cook (1975). These estimates are derived from the graph of his cost curve.

turing the product they analyzed. When the basic metal cost index and the other explanatory variables are included in the cross-sectional analysis (Case A in the table), the influence of the unit cost elasticity is lessened, and the estimated elasticity is smaller than in all of the other cases. For the one case where there are other explanatory variables to measure the influence of the nature of the processing requirements, the magnitude of the expected value of the output elasticity is smallest. Why might this be so? Across the metalworking industries, the unit cost of the basic metals purchased tends to decrease as the number of pounds of metal processed per establishment increases. Low-volume producers tend to use more expensive basic metal inputs, and high-volume producers tend to use lower cost, more standardized inputs. One explanation for the significance of the decrease in unit basic metal cost with increasing output is that lower cost, more standardized material inputs are suggestive of simplified and standardized product designs as well as simplified and standardized processing requirements. This implies that increasing the scale of production means more than just increasing the volume of production. The regression equation which includes a variable representing the nature of the processing requirements in addition to an output variable gives a smaller estimate of the magnitude of the output elasticity than regression equations which include only the output variable. Without an explanatory variable to measure the influence of changes in the types of material used and in the way the product is made, the output elasticity incorporates the effects of increasing batch size *and* of altering the nature of the processing requirements. The presence of such an explanatory variable in the model, however, measures some of the effects of altering the material inputs used, and the output elasticity picks up only the effects of increasing batch size. This would reduce the influence and, therefore, the magnitude of the output elasticity.

The unit cost elasticity of –0.295, derived from the regression relation which includes both output and complexity parameters, is used to approximate a lower bound on the decrease in unit cost that would be realized in a high-volume batch-production factory. The decrease in unit cost given by this elasticity estimate is viewed as a lower bound because it assumes that the level of output is increased without altering the types of material inputs. Using this elasticity, it is assumed that even though the plant is producing a high volume of output across all products, each product type is not designed to minimize the complexity of processing requirements, as is typically the case with specialized products in a conventional batch-production factory.

The unit cost elasticity of –0.436, derived from the regression relation which includes only the output parameter, is used to approximate an upper bound on the percentage of decrease in unit cost that would be realized. This unit cost elasticity incorporates both the effect of increasing the quantity of output produced as well as the effect of using more standardized materials. It is assumed

Table 4-20: The Elasticity of Unit Cost Components with Respect to Output

| | | Estimated Elasticity of Unit Cost Component With Respect to Output | |
Unit Cost Component	Average Share of Unit Cost[a] (%)	Scale + Complexity Variables[b]	Scale Variable Only[c]
Total value added[d]	100.0	−0.295	−0.436
Labor value added[e]	55.9	−0.345	−0.461
Production worker costs	35.4	−0.306	−0.440
Salary labor costs	20.4	−0.406	−0.536
Nonlabor value added	44.1	−0.232	−0.408

[a]Average share of total cost for 101 metalworking industries included in sample.
[b]Output elasticity for each cost component is the estimate of b_1 in
$$E[\ln(\text{unit cost component})] = b_0 + b_1\ln(m/e) + b_2\ln(bmci)$$
$$+ b_3\ln(1 + pmbm) + b_4\ln(w) + b_5\ln(c)$$
See Appendix E for an explanation of the independent variables used in the regression analysis.
[c]Output elasticity for each cost component is the estimate of b_1 in
$$E[\ln(\text{unit cost component})] = b_0 + b_1\ln(m/e)$$
[d]Value added = labor value added + nonlabor value added.
[e]Labor value added = production worker costs + salary labor costs. (Does not add to total shown because of rounding.)

that even though batches of specialized products are being made, each product is optimally designed for ease of manufacture, as is typically the case with standardized products made in a conventional mass-production factory.

Determining why value added per unit decreases as the level of output increases across metalworking industries with 1977 technologies is useful for understanding the limitations of this analysis. Total value added can be broken down into the primary components of labor value and nonlabor value added. Labor value can be further subdivided into production worker costs and salary labor costs. Table 4-20 shows how value added is proportioned among these components for the 101 metalworking industries in the sample. An estimate of the elasticity of each unit cost component with respect to output derived from the expanded regression model is also shown. As the level of output increases, unit labor costs decrease more than unit capital costs. As the level of output increases, unit salary labor costs decrease more than unit production worker costs.[7] The elasticity of total value added per unit is a weighted average of the elasticity of labor value added per unit and nonlabor value added per unit, where the weights are given by the cost component's average share of total value added.

These estimated elasticities for total value added and for the unit cost components are derived from relationships between unit cost and output as of 1977. The analysis assumes the use of conventional technologies (e.g., manually auto-

7. Reasons for the decrease in salary labor costs are given in Chapter 3.

mated or semiautomated machines for low-volume production and special-purpose machines for high-volume production). An issue of concern is whether these relationships between unit cost and output when based on the use of conventional technologies will closely approximate the relationships between unit cost and output when using the new types of flexible technology. One might argue that if new types of flexible technology were used as opposed to conventional types, both unit labor costs and unit capital costs would decrease more sharply as the level of output increases. Such an argument is supported by descriptions of the flexible manufacturing systems given earlier. Management of both Fanuc and Yamazaki claims that it would have taken substantially more capital investment and labor to increase output by the same amount if more conventional types of technology had been used. If this were the case it would mean that, with the complexity of the metal-shaping activities held constant, the magnitude of the unit cost elasticity would be greater than -0.295. How much greater is not known. Had there been enough information available to systematically estimate the relationships between unit cost and output in a plant employing the new types of flexible automation, I would have used this information directly instead of trying to infer the relationship from the analysis of unit cost across metalworking industries. I acknowledge the possibility that the use of new types of flexible automation could cause unit capital cost and/or unit labor cost to decrease more rapidly with an increase in output than what is implied from the analysis of 1977 technologies, but I have not analyzed this possibility. It is another reason for considering a range of values for unit cost elasticity with respect to output, as I have done here.

Table 4-21 presents the estimated percentage of reduction in unit cost that would result from increasing output by 50–1000 percent in a batch-production plant. For a given increase in output, the decrease in unit cost is calculated using a low and high value of the unit cost elasticity. If output were to increase by 100 percent, the estimated decrease in unit cost would range from 18–26 percent. If output were to increase by 500 percent, the estimated decrease would range from 41–54 percent. If output were to increase by 1000 percent, the estimated decrease in unit cost would range from 50–65 percent. As a point of comparison, it was noted earlier that Fanuc management claims that the total cost of producing servo motors would be cut by 30 percent as a result of building a new flexibly automated batch-production facility.

4.7. The Plant Consolidation Scenario

If a flexibly automated plant could be built with several times the capacity of a conventional batch-production plant, then several old plants could be closed down and their production consolidated into the new facility, which would have the flexibility to produce a mix of different products. This seems to be a likely

Table 4-21: Decrease in Unit Cost Derived from Estimated Output Elasticity

Percent Increase in Output	Percent Decrease in Unit Cost Assuming Elasticity Equals:	
	−0.295[a]	−0.436[b]
50	11.3	16.2
100	18.5	26.1
200	27.7	38.1
300	33.6	45.4
400	37.8	50.4
500	41.0	54.2
1000	50.7	64.8

[a]Output elasticity derived from estimate of b_1 in
$$E[\ln(va/m)] = b_0 + b_1\ln(m/e) + b_2\ln(bmci) + b_3\ln(1 + pmbm) + b_4\ln(w) + b_5\ln(c)$$
[b]Output elasticity derived from estimate of b_1 in
$$E[\ln(va/m)] = b_0 + b_1\ln(m/e)$$
$$\Delta \text{ unit cost} = (1 + \Delta\text{output})^{-b_1} - 1$$
$$b_1 = \text{elasticity of unit cost with respect to output}$$

scenario if the flexibly automated plant were built in a mature industry where the potential for market growth is limited.

To understand the role of flexibility in the plant consolidation scenario, consider a company with three plants where each plant makes a different type of widget. Suppose one of the three plants is converted to a flexibly automated factory and is now capable of meeting the total demand for all three types of widgets. The flexible widget plant can now produce the two other widgets in addition to its own widget at a price below that of the remaining two conventional producers. Higher volumes and lower unit costs are achieved in the flexibly automated plant, even though it is now producing a more varied product line than before. As a result, the two other plants are shut down, and the company's total output is produced in the one flexibly automated plant.

If each of the three old plants employed P_0 workers, the base level of employment equals $3P_0$. The new level of employment in the one flexibly automated plant, P_*, is estimated from the equation:

$$P_* = P_0(1 - \Delta a_L)(1 + \Delta Y) \tag{4.3}$$

where

$-\Delta a_L$ = percentage of decrease in unit production labor requirement
    = −0.306 to −0.440

$\Delta Y = 2$, or a 200 percent increase in output

The decrease in the unit labor requirement, $-\Delta a_L$, is taken from the estimates of the elasticity of production labor cost with respect to output shown in Table 4-20. Given these values for the parameters in Equation (4.3), the new employment level, P_*, ranges from

$P_* = 1.68P_0$ where $-\Delta a_L = -0.440$

to

$P_* = 2.08P_0$ where $-\Delta a_L = -0.306$

The one flexibly automated plant would employ 31–44 percent fewer workers than the three conventional plants. While this is a substantial decrease in labor requirements, it could well be a conservative estimate. Descriptions of FMS applications indicate that even fewer workers would be required to produce several times the output of a conventionally organized plant. If this were the case, the new employment level in the high-volume plant would be less than the base employment level in one of the old plants, and labor requirements would decrease by 66 percent or more. Using the elasticities of unit cost with respect to output shown in Table 4-21, it is estimated that unit cost would decrease by 28–38 percent if the output of one flexibly automated plant were 200 percent more than one of the old plants. Thus, not only would there be a substantial decrease in labor requirements within an industry if older facilities were shut down and production were consolidated into flexibly automated facilities, there would also be large economic benefits due to a substantial decreases in production costs.

In Chapter 3 it was estimated that most metalworking industries are dominated by batch production. Therefore, if it were possible to build flexibly automated factories and to consolidate the production of several plants into one facility, there could be very dramatic changes in labor requirements and in the industrial structure within the metalworking industries. Given this possibility, it is important to give a more detailed consideration to projecting the rate at which flexibly automated factories will be built. It is also important to give a more detailed consideration to the potential for consolidating older plants into flexibly automated facilities in order to evaluate whether or not there might be a dramatic change in production labor requirements over the next few decades.

5

Conclusion

The discussion in Chapter 2 of the impact of industrial robots on future job displacement illustrates the difficulties of conducting such an analysis. The focus of the chapter is on forecasting the number of workers that will be displaced by robots. Yet, labor requirements on the factory floor are being affected by a wide range of automation technologies in addition to robotic manipulators, such as computer-controlled machine tools, material-handling systems, and programmable controls. Combinations of these other types of programmable automation can sometimes be substituted for robotic manipulators. For example, there are situations where a material-handling system with the appropriate type of programmable controls can perform the functions of a material-handling robot. There are also cases where the boundaries between what is a robot and what is some other form of programmable automation are not so clear. And as automation in general becomes "more robotic" (i.e., more programmable, more adaptable, and more multifunctional), the distinctions between what is and what is not an industrial robot will become increasingly blurred.

Clearly, the question of how many workers are likely to be displaced by industrial robots in future years is only one element of the more general question of how many workers in manufacturing are likely to be displaced by factory automation. Forecasting technologically induced displacement is further complicated by the fact that changes in production technology are increasingly accompanied by efforts to simplify or rationalize both the design of the product and the sequence of production activities. These types of change also affect the labor requirement per unit of output. It is evident that multiple forms of technology and multiple factors are altering production labor requirements per unit of output on the factory floor. Given that, one wonders if it is possible to forecast the component of job displacement that is attributable specifically to industrial robots. In some cases, perhaps. In many cases, undoubtedly not. And even if the portion of job displacement induced by robot use could be disentangled from overall technologically induced displacement, is it important to do this? Probably not.

Why, then, is Chapter 2 full of material which attempts to estimate the job displacement impact of robot use? The answer lies in the mind-set that seemed to prevail among both technologists and analysts of technological change in the late 1970s and early 1980s. That was the time when an understanding of the inefficiencies of current manufacturing methods had become a popular "cutting edge" topic for a certain set of the technological and managerial avant-garde. It was also the period in which the ground swell of both commercial and research interests in robotics began. In 1980, Joseph Engelberger, recognized as the father of the commercial robotics industry, released his book *Robotics in Practice,* which spurred interest in robot applications. At that time, there was publicity around the fact that Japan was then leading the world in the number of robots installed in factories. The first commercially available robot truly controlled by computer software, the Cincinnati Milacron T3, had recently appeared on the market, demonstrating that new technology could improve robot capabilities. Much publicity surrounded the inauguration of new research centers, such as the Carnegie Mellon Robotics Institute, that set out to develop a new generation of "intelligent robots."

Although it was clear back then that robots were not the only new technology relevant for automating factory operations, it seemed that they were the key technology. Robot use was regarded as the bellwether of a great change that was on the threshold of taking place in factories throughout the world. Robots were going to have an important impact! The issues then, became figuring out what that impact would be. It was important to think ahead and anticipate some of the consequences of the inevitable "robotization" of the factory.

Given the mind-set that existed in the late 1970s and early 1980s, if one were going to think about the impact of technology on factory operations, it seemed reasonable (as well as fashionable) to think about changes that would occur as a result of using robots. One obvious consequence would be that robots would perform some of the tasks currently performed by production workers. Thus, as part of looking at the impending impact of robot use, what seemed important, perhaps critical, was to estimate how many jobs would likely be taken over by robots. This is why several of the studies reviewed in Chapter 2 specifically address the issue of how many workers would likely be displaced by robots over the next 10–20 years.

What are the implications of the points made above? One implication is, that if the objective of analysis is to predict future levels of technologically induced displacement of production workers, it does not make sense to scrutinize or fine tune many of the details of the robot displacement studies presented in Chapter

2. Details on the number of robots projected to be in use are important to the issue of forecasting the size of the robot market. But such details, even if correct, would provide only part of the information required to predict future levels of worker displacement in manufacturing. The same point holds for details on the fraction of jobs within specific occupational categories that could potentially be performed by robots. Such details would be important for forecasting displacement if one could obtain estimates that represent the effect of automation and technological change in general, as opposed to the effect of robotic manipulators in particular.

For policy analysts and planners interested in manufacturing, anticipating the likely magnitude and timing of job displacement on the factory floor has always been important. It still is. What is evident from the experience of manufacturers in the 1980s is that displacement is being induced by multiple types of change in the production process, as well as by change in managerial practices and in the design of products. Therefore, if one sets out to forecast displacement of production workers within an industry or sector, it should be approached from a broader perspective that encompasses the full set of factors affecting labor requirements. It should not be approached from a technology-specific perspective, such as trying to forecast the number of workers that will be displaced by robots.

Even if we are clear about the type of worker that could be displaced by programmable automation in general or by robots in particular, there is another issue that complicates the forecasting of displacement. We are not certain about when, or even whether, the technology will actually be adopted by potential users. The ability to predict the full cost and benefit of adopting the technology is key to forecasting its rate of diffusion, and hence it is key to estimating the rate at which displacement will occur. The problem is, the *total* costs and benefits of implementing and using new technologies on the factory floor are still not fully understood. This leads to uncertainty in the analysis of whether or not potential users will have a strong enough economic incentive to adopt new technology. Because of such uncertainty, there will continue to be ambiguity as to the size of the gap between engineering-based estimates of the theoretical potential for displacing workers and more economic-based estimates of the number of workers that may actually be displaced within a given time frame.

Within the past several years, the topic of financial justification for new technologies in manufacturing has received much attention in the communities of both the practitioner and the researcher. As a result of these efforts, there is growing understanding of the economic impact of using new technologies. The major contribution of the new work on financial justification has been in recognizing and quantifying a fuller range of financial benefits beyond reduction of direct labor cost (e.g., financial benefits associated with changes in levels of quality, inventory, productivity, flexibility, innovation). What has received less attention, though, is the recognition and quantification of the full range of costs for planning, implementing, and operating robots and other new factory-floor technologies. The more time one spends around automation projects in industry, especially large projects, the more evident it becomes that the cost of the equipment is only a small fraction of the cost of automating. For example, there are costs of focusing large blocks of managerial time on automation projects, training costs, support costs, and many costs associated with making transitions within the organization. And, as pointed out in the analysis in Chapter 2, these costs can vary over time as more experience is gained with using the technology. As the full range and dynamic nature of costs and benefits of using technologies are recognized and as a consensus emerges on approaches for quantification, the ability to forecast the rate of diffusion should improve. This in turn would improve our ability to distinguish between predictions of upper bounds on potential displacement and predictions of what will most likely happen in the near-term future.

In addition to these difficulties in forecasting technologically induced displacement, there are limitations on what can be learned from such an analysis. Some of these limitations were reviewed in the section "Perspective on Findings," at the end of Chapter 2. The first limitation is that even if we know for sure the number of jobs in an industry that could be performed by robots or other machines, we do not know how many of the workers who do these jobs will become redundant. This depends on the total demand for the industry's output. Thus, while the Leontief and Duchin (1984) study forecasts that several million jobs will be performed by robots and other types of factory-automation technology over a 20-year period, it also forecasts there will actually be more people working in these same jobs because the forecasted growth in the level of output would offset the effects of displacement. Technologically induced displacement is only one of the factors effecting changes in the level of employment within occupations in industry. And the consensus of analysts is that, compared with the effect of changes in the level of demand, displacement is not even a primary factor affecting employment levels. The second limitation is that the most dramatic labor-force impact resulting from the use of new factory automation technologies appears to be changes in the nature of work, and hence in skill requirements for the related workers.

On one hand, an industrywide or sectorwide study of displacement is too narrow. It does not address the issue of net changes in employment within occupational groups because the influences of macroeconomic factors, such as demand levels, are not considered. Economic models of industries and of the economy are needed for this. On the other hand, a study of displacement alone is too broad, because it does not provide enough detail on how the use of the technology will change the nature of the work and the skill requirements. Organizational-level studies of technology implementation and use are needed for this.

Despite the difficulties and limitations of analyzing the impact of robot use on job displacement, the studies on this topic reviewed in Chapter 2 have still made an important contribution to the more general analysis of the impact of robotics and factory automation on the labor force. These studies have been used as a starting point, a point of contrast, or a general motivator for a number of other studies which have subsequently been carried out by other researchers. For example, the study by the Office of Technology Assessment (OTA, 1984), which looks more broadly at the labor-force impact of factory automation, used as its starting point several of the studies on the impact of robot use. The Leontief and Duchin (1984) study, which simultaneously considers the effects of automation technologies and macroeconomic influences on job requirements, also used several of the robot displacement studies as a data source for making assumptions about the impact of technology. Also, studies by organizational researchers, which have looked with more depth into how the use of factory automation is affecting the nature of work within specific companies, cite the robotic displacement studies as motivation for the widespread importance of researching the topic. The studies reviewed in Chapter 2 are still frequently referred to in more recent work in the area. For this reason, it is useful to have the material compiled in one place, as is done here, even though much of the analysis was completed several years ago.

In Chapter 3, two important premises about the nature of discrete-parts manufacturing are examined: (1) most of the value added in metalworking is generated by batch production, and (2) there is a large differential in unit cost between customized goods produced in small batches and standardized goods that are mass produced. Nearly all in the avalanche of articles appearing in the last decade that have argued the importance of robotics and flexible forms of automation make reference to these two premises. These points are so widely accepted within the manufacturing management and engineering communities that they are now taken on faith, and it is typically not even necessary to substantiate them. For example, the claim that most of the value added in metalworking is generated by batch production is typically supported with anecdotes from and references to studies done "long ago." Not surprisingly, the estimates of the distribution of value added by mode of production and of the cost differential between batch and mass production reported here corroborate the "conventional wisdom." Nonetheless, it is still important to provide empirical support for such widely made claims, especially since they are so central to the argument for accelerating the use of robotics.

The following scenario was analyzed in Chapter 4: a conventional factory (circa 1980) which produces batches of specialized products is reorganized and automated with robots and other types of programmable automation technology so that it can produce specialized products in a continuous fashion around the clock. The framework used in Chapter 4 for estimating the resulting reduction in unit cost is the assumption that this "ideal" flexibly automated plant, capable of continuously producing specialized products in small batches, would have some of the operating characteristics of conventional mass-production plants which produce standardized products in large volumes. Relationships between unit cost and the level of output derived from a cross-sectional analysis of industries using the 1980's generation of production technology provide the basis for estimating potential cost reductions in the hypothetical high-volume batch-production facility. The result of the analysis is that severalfold increases in the output of a batch-production facility would lead to a substantial decrease in unit cost. For example, if the output of the new facility were twice that of the conventional one, the estimated decrease in unit cost would range from 18–26 percent. If it were 10 times greater, the estimated decrease in unit cost would range from 50–65 percent.

Because of the inferences and approximations required to do the work, the analysis is only suggestive of the economics of production in a newly designed, flexibly automated plant. Nonetheless, if the inference of this analysis is correct, and it is the case that a flexibly automated batch-production plant would have a substantial cost advantage over a conventionally organized facility, one would expect that this type of plant would begin diffusing throughout manufacturing industries. In fact, the examples of the flexible manufacturing systems for machining discussed in Chapter 4, particularly the Japanese systems, are the initial attempts at building such facilities. And many major manufacturers throughout the world have plans in progress to construct flexibly automated plants for batch production.

If a plant could be built having several times the capacity of a conventional batch-production plant and the flexibility to produce a mix of different products, it would be possible to close down several existing plants and consolidate their production into the new facility. This seems likely if the flexibly automated plant were to be built in a mature industry where the potential for market growth is limited. An example worked out in Chapter 4 shows that if three plants were closed down and their output consolidated into one high-volume, flexibly automated plant, total labor requirements would decrease by 30–40 percent. This estimated decrease in labor requirements is based on the elasticity of unit production-labor cost, which is estimated from the regression analysis based on the use of conventional types of technology across low-, medium-, and high-volume industries as of 1977. More recent information available from the existing flexible manufacturing systems in operation indicates that the one flexibly automated plant would most likely have substantially fewer workers than even one of the smaller plants it replaces. If this were the case, the percentage of decrease in total labor requirements would be much larger.

Since the flexible factory scenario holds the largest promise for reducing unit cost and potentially poses the largest threat to employment in an industry, it warrants more extensive analysis. Further research should focus on a more refined and direct analysis of the economics of production in flexibly automated factories, and on forecasts of their use within specific industries.

Appendices
Bibliography
Index

Appendix A
The Standard Industrial
Classification (SIC) System

What Is an Industry?

It is necessary to adopt a standard for defining separate "industries" within manufacturing. The standard industrial classification (SIC) system is used to define industries and products throughout this volume. This is the system of industrial classification that has been developed over a period of many years under the guidance of the Office of Federal Statistical Policy and Standards with the U.S. Department of Commerce. All data collected by the Bureau of the Census as part of the economic census, including the *Census of Manufactures,* is organized according to the SIC system. The following overview of the classification system is excerpted from the introduction of the *1977 Census of Manufactures General Summary* (Bureau of the Census, 1981a):

The SIC system was developed to classify establishments as distinguished from similar systems used to classify companies or enterprises. An establishment is defined as a single physical location engaged in one of the industry categories of the SIC.

The SIC coding system is designed to describe industries and operates in such a way that the definitions become progressively narrower with successive additions of digits. For manufacturing, there are 20 very broad 2-digit SIC Major Groups (20 to 39) which are subdivided into 144 3-digit SIC Groups (201 to 399) and into 452 4-digit industries (2011 to 3999). Based on the SIC Manual, the Bureau of the Census has developed a product coding system along similar lines which contains about 1,500 5-digit product classes and approximately 13,000 individual 7-digit products . . . . The 7-digit products and 5-digit product classes are considered the primary products of the industry with the same first four digits as the product code. . . . The 7-digit product code number 2023212 refers to canned evaporated milk. The first five digits of this code number, 20232, refer to the "product class" canned milk, and the first four digits, 2023, refer to the industry in which these products are made (the condensed and evaporated milk industry).

An establishment is classified in a particular industry if its production of the primary products of that industry exceeds in value its production of products of any other single industry. The industry code assigned to the establishment is derived from a summation of values for 7-digit product codes to their 4-digit industry and the selection of the largest 4-digit value as the establishment's industry classification. However, . . . the industry classification of an establishment may be determined not only by the products it makes but also by the processes employed or materials used in making those products.

The extent to which industry and product statistics may be matched with each other is measured by two ratios. The first of the ratios, called the primary product specialization ratio, is the proportion of product shipments (both primary and secondary) of the industry made up of primary products. The second, defined as the coverage ratio, is the proportion of primary products shipped by the establishments classified in the industry to total shipments of such products by all manufacturing establishments.

It is important that the reader be familiar with the hierarchical nature of the SIC system. All three-digit groups of industries (e.g., SIC 341, metal cans and shipping containers) belong to the major group identified by the first two of the three digits (e.g., SIC 34, Fabricated Metal Products). Similarly, all four-digit industries (e.g., SIC 3411, metal cans, and SIC 3412, metal shipping containers) belong to the group identified by the first three digits (SIC 341). The Bureau of the Census reserves the term *industry* for the four-digit level of aggregation, the term *group* for the three-digit level, and the term *major group* for the two-digit level. It is also important to note that an industry is named for its primary product, but that other secondary products are also made within the industry.

Appendix B
Revised Survey Data on the Percentage of Workers, by Occupation, Whose Jobs Could Be Replaced by Robots

During the first six months of 1981, members of the Robot Institute of America (RIA) were surveyed to estimate the percentage of jobs within various occupations that could potentially be robotized. The 24 respondents who answered this part of the survey were either experienced robot users, or were in the process of evaluating robot applications.[1] The responding firms are listed in Table B-1. The survey was conducted anonymously. Although, most of the respondents identified themselves and their names are listed in Table B-1. However, the data from each individual respondent have been kept confidential. The number of production workers and the average batch size within each establishment that responded are also indicated. In several cases, the respondents worked for the corporate headquarters of manufacturing companies with multiple establishments. For these cases, the figures shown in Table B-1 for the number of production workers and the average batch size are typical of the company's establishments.

The RIA members were asked to estimate what percentage of jobs within a given occupation could be done by Level I and Level II robots. In the survey, a Level I robot was defined as an insensate machine similar to those on the market in 1981. Level II robots were broadly defined as "the next generation of robots with rudimentary sensory capabilities." More comprehensive definitions of the

1. Preliminary versions of these survey results have previously been published in Carnegie-Mellon University 1981 and in Ayres and Miller, 1983. Estimates of potential substitution by occupation composed only part of the original robotics survey, and many respondents omitted this part while answering the other parts of the survey. In total, 52 establishments were surveyed and/or interviewed. The full sample of 52 respondents accounted for 1291 robots as of January, 1981, which was over a third of the estimated total U.S. robot population at the time.

Table B-1: Firms Responding to the CMU Survey on the Potential for Robotizaton in Selected Occupations

Firm Number	Firm Name	Number of Production Workers	Batch Size[a]
1	ALCOA[b]	> 1000	mass
2	Stanley Tools	> 1000	mass
3	Acme Wire Stamping	1–99	large batch
4	Air Products and Chemicals	100–499	custom
5	Hatch Associates	100–499	mass
6	Bridgeport Machines	> 1000	medium batch
7	Kysor Machine Tool	1–99	large batch
8	Hall Industries	1–99	large batch[c]
9	Whirlpool	> 1000[d]	mass[e]
10	General Electric	> 1000	custom medium batch[f]
11	Westinghouse[g]	> 1000	custom (80%)
12	Union Switch and Signal	500–999	medium batch
13	Ford Motor Company	> 1000	mass
14	Caterpillar Tractor	> 1000	large batch
15	Lockheed-Georgia Co.	> 1000	custom
16	Rockwell International Automotive Division	> 1000	mass
17	Hoover Vacuum Cleaner	> 1000	large batch
18	General Dynamics	> 1000	medium batch
19	Packard Electric–GM	> 1000	mass[h]
20	ALCOA Forging	500–999	medium batch[i]
21	Home appliance producer[j]	> 1000	mass
22	Auto-related company[k]	> 1000	medium batch
23	Unknown	100–499	large batch[l]
24	Unknown	100–499	mass

The survey was conducted anonymously, but the respondents who are named here identified themselves. However, the individual survey data from each respondent have been kept confidential.

[a]Batch size definitions:

custom	1–100 units per batch
medium batch	101–1000 units per batch
large batch	1001–10,000 units per batch
mass	> 10,000 units per batch

Custom and medium-batch producers are aggregated to form the category "small batch" producers in later tables.

[b]ALCOA is primarily a mass producer of aluminum. They also have establishments which fabricate aluminum products in batches (see firm number 20). The corporation as a whole is classified as a mass producer. The establishment making aluminum forgings, which responded separately, is classified as a medium-batch producer.

[c]Hall Industries produce screw machine products. The value of output is distributed as follows:
80% large batch   20% custom-small batch

[d]The number of production workers in an establishment is inferred as follows. Household laundry equipment is the primary product of SIC 3633. In this industry, 7.4/15.9 = 47 percent of production workers are employed in establishments which have 500–2499 production workers. This combines two size classes, 500–999 workers and 1000–2499 workers. In this industry, 43 percent of production workers are employed in establishments with 2500 or more employees. Establishments in two size classes, 500–999 workers and 1000–2499 workers, account for 47 percent of the production work force. Assuming that more than 4 percent of the workers in these two size classes are employed in establishments in the larger size class, most workers are employed in establishments with 1000 or more employees. Thus, the size class for Whirlpool is estimated to be 1000 or more.

(continued on the following page)

Table B-1: Firms Responding to the CMU Survey on the Potential for Robotizaton in Selected Occupations (*continued*)

eWhirlpool manufactures household laundry equipment and is located in SIC 3633, the house laundry equipment industry. This industry is classified as being dominated by mass production in Chapter 3. This particular establishment, which belongs to one of the major producers in the industry, is classified as a mass producer.

fBatch size distribution is as follows:

50% 1–100   50% 101–1000

gResults are averaged for whole corporation.

hBatch size is distributed as follows:

84% > 10,000   10% 1,001–10,000   5% 101–1000   1% 1–100

iBatch size is distributed as follows:

40% 1001–10,000   40% 101–1000   20% 1–100

jName of company unknown. Number of production workers in plant is estimated. Batch size is distributed as follows:

85% > 10,000   15% 101–10,000

kName of company unknown. Batch size is distributed as follows:

5% > 10,000   25% 1001–10,000   60% 101–1000   10% 1–100

lName of company unknown. Batch size distribution is as follows:

50% > 10,000   40% 1001–10,000   10% 101–1000

The only way in which a Level I robot can respond to the external environment is through predetermined alternative sets of instructions where all aspects of the alternative tasks are explicitly specified in advance. Predetermined alternatives may be invoked via limit switches, timing devices, or simple transducers which transmit binary inputs or measure predetermined thresholds.

Level II robot.—A robot system where the goals *can* be modified through the result of adaptive feedback and control processes which sense the external environment via some type of transducer. Since there is feedback between the robot and its external environment, it is possible to perform a task without having to explicitly specify all of its aspects in advance.

The distinguishing feature of a Level II robot is that it can dynamically alter its plans in response to changes in the external environment.

The survey results are summarized in Table B-2.[2] For each occupation included in the survey, the number of responses, the minimum response, the maximum response and the average response are shown for Level I and Level II robot technologies. The average responses are computed in three different two levels of robot technology are given below, although these definitions were not included in the survey form.

Level I robot.—A robot system where the goals *cannot* be modified through the result of feedback and control processes that sense the external environment via some type of transducer. Since there is no feedback between the robot and its external environment, *all* aspects of the task must be explicitly specified in advance of performing it.

2. The estimates published here of the percentage of workers by occupation that could be displaced by robots may not match the previously published survey results for two reasons. The survey results shown here include the responses of eight additional users, and the average response for each occupation has been calculated in a different manner.

Table B-2: Summary of Revised Responses of the Percentage of Jobs That Could Be Robotized, by Occupation and by Level of Robot Technology

Occupation	Level of Robot Tech.	Number of Response	Minimum Response	Maximum Response	Average, Simple	Average, Weighted by Distribution of Employees	Average Weighted by Batch-Size Distribution
Dip plater	I	6	20	100	48.3	55.7	43.7
	II	6	50	100	78.3	79.7	81.5
Punch press operator	I	5	10	100	45.0	44.3	39.0
	II	5	60	100	76.0	75.0	67.8
Painter	I	16	0	100	40.0	43.5	37.7
	II	15	0	100	62.3	66.8	60.5
Riveter	I	3	5	100	38.3	40.5	25.2
	II	3	10	100	50.0	51.8	35.9
Shotblaster/	I	6	10	100	35.8	35.6	31.9
sandblaster	II	6	10	100	35.8	35.6	31.9
Drill press operator	I	5	25	50	33.0	32.5	30.1
	II	5	60	75	67.0	67.0	64.8
Etcher-engraver	I	5	0	100	27.0	29.9	24.3
	II	5	0	100	53.0	59.2	40.3
Welder	I	17	0	60	23.8	25.5	22.0
	II	17	10	90	45.6	45.7	47.8
Coil winder	I	7	0	40	23.6	24.5	24.8
	II	7	15	50	38.6	40.2	39.7
Heat treater	I	3	5	50	21.7	22.7	16.8
	II	3	40	90	60.0	61.1	52.5
Machine tool	I	20	0	90	19.8	21.7	18.4
operator, NC	II	19	0	100	44.7	46.5	41.2
Grinding/abrading	I	5	10	20	18.0	18.2	19.3
machine operator	II	5	30	100	58.0	57.5	53.5
Lathe/turning	I	5	10	20	18.0	18.2	19.3
machine operator	II	5	25	65	50.0	50.4	50.0
Conveyor operator	I	14	0	50	17.5	14.9	18.7
	II	14	15	65	33.2	41.9	33.2
Electroplater	I	6	5	40	17.5	18.1	15.2
	II	5	15	60	43.0	42.9	44.5
Milling/planing	I	5	10	20	16.0	16.1	16.9
machine operator	II	5	40	60	52.0	52.1	50.7
Filer/grinder/buffer	I	13	0	35	12.1	9.8	11.6
	II	13	5	75	27.7	27.6	26.2
Packager	I	15	0	40	11.8	10.8	8.7
	II	15	0	70	27.1	26.5	20.5
Pourer	I	3	5	20	11.7	10.9	13.1
	II	3	10	30	20.0	21.4	20.0
Assembler	I	19	0	40	10.3	8.9	9.5
	II	19	15	60	31.1	28.8	29.4
Composites and	I	1	10	10	10.0	10.0	10.0
bonded structures worker	II	1	40	40	40.0	40.0	40.0
Sheet metal	I	1	10	10	10.0	10.0	10.0
operator	II	1	40	40	40.0	40.0	40.0
Inspector	I	19	0	25	8.2	7.5	7.9
	II	19	5	60	29.2	30.4	28.3

(continued on the following page)

Table B-2: Summary of Revised Responses of the Percentage of Jobs That Could Be Robotized, by Occupation and by Level of Robot Technology (*continued*)

Occupation	Level of Robot Tech.	Number of Response	Minimum Response	Maximum Response	Average, Simple	Average, Weighted by Distribution of Employees	Average Weighted by Batch-Size Distribution
Caster	I	4	5	15	7.5	8.6	7.2
	II	3	10	20	15.0	15.2	15.9
Electronic wirer	I	3	0	10	6.7	7.0	7.3
	II	3	10	50	30.0	27.6	32.0
Order filler	I	9	0	20	6.7	6.9	5.5
	II	9	0	80	29.4	31.7	25.3
Tester	I	17	0	10	5.8	4.8	5.1
	II	17	0	30	11.4	11.3	10.8
Mixer	I	3	0	10	5.0	6.0	5.2
	II	3	10	10	10.0	10.0	10.0
Tender	I	2	0	10	5.0	6.4	5.5
	II	2	20	20	20.0	20.0	20.0
Millwright	I	4	0	15	3.7	4.4	3.0
	II	4	0	15	3.7	4.4	3.0
Kiln furnace	I	3	0	10	3.3	2.9	2.0
operator	II	3	5	20	13.3	14.7	10.5
Tool and	I	8	0	5	1.5	1.3	0.9
die maker	II	8	0	60	16.6	15.3	9.2
Oiler	I	3	0	0	0.0	0.0	0.0
	II	3	0	0	0.0	0.0	0.0
Rigger	I	2	0	0	0.0	0.0	0.0
	II	2	0	0	0.0	0.0	0.0
Trader/helper	I	1	0	0	0.0	0.0	0.0
	II	1	50	50	50.0	50.0	50.0

Occupations are ordered by average (simple) response for Level 1.

ways. The first technique is to compute a simple average where the estimates from each respondent are weighted equally. For each occupation, the simple average is given by

$$a_{simple} = (1/n) \sum_{i=1}^{n} p_i \qquad (B.1)$$

where

n = total number of respondents for the specified occupation

p_i = the ith respondent's estimate of the percentage of workers in the given occupation whose jobs could be performed by a robot (Level I or Level II)

A second technique is to group the estimates for a given occupation and level by the number of production workers in the responding establishment. Within each of the different size classes the responses are averaged. The simple averages for each size class are combined by weighting factors which are derived from the percentage of production workers in metalworking employed in that

Table B-3: Distribution of Metalworking Production Workers by Size of Establishment and Major Group, 1977

Industry	Number of Production Worker Employed (in millions)	Percent of workers employed in establishments with this many workers:			
		1–99	100–499	500–999	≥ 1000
SIC 34	1191.6	35.0	37.8	11.8	15.5
SIC 35	1413.8	30.0	26.5	14.2	29.3
SIC 36	1191.4	12.9	28.8	17.6	40.7
SIC 37	1284.4	8.0	14.5	7.7	69.8
Total, SIC 34–37	5082.1	21.6	26.7	12.8	38.9
Total, manufacturing	13,691.0	26.1	34.5	13.5	25.9

Source: 1977 Census of Manufactures (Bureau of the Census, 1981c,d).

size class. The weighting factors used are taken from the bottom of Table B-3, which shows the percentage employment of production workers in SIC 34–37 by size of establishment (where size is defined as the number of production workers employed in the establishment). For each occupation, the average, weighted by size of establishment, is given by

$$a_{size} = \sum_{j=1}^{4} s_j a_{simple,j} \qquad (B.2)$$

where

j = index of size classes of respondents

$j = 1$, 1–99 production workers in the establishment

$j = 2$, 100–499 production workers

$j = 3$, 500–999 production workers

$j = 4$, > 1000 production workers

s_j = percentage of metalworking production workers employed in establishments of given size class. For example, $s_1 = 0.216$ = percentage of metalworking production workers employed in establishment with from 1 to 99 production workers.

$a_{simple,j}$ = simple average of substitution estimates of respondents in size class j.

A third technique, also a weighted average approach, is to group the estimates for a given occupation and level by the batch size of the responding establishment. Within each of the different batch-size ranges, the responses are averaged. These averages are combined by weighting factors which are derived from an estimate of the percentage of value added that is generated within each batch-size range. Estimates of percentage of value added generated by industries dominated by small- and large-batch and mass production are shown in

Table B-4: Estimated Percentage Distribution of Value Added in Metalworking by Mode of Production, 1977

Mode of Production	Value Added
Small batch[a]	57.9
Large batch	16.4
Mass	25.7

Source: Chapter 3.

  [a]In Chapter 3 custom and small batch production is estimated to account for 31.5 percent of the value added in SIC 34–37, and medium batch production is estimated to account for 26.4 percent. Custom, small batch, and medium-batch are aggregated here to form the category "small batch."

Table B-4. For each occupation, the average weighted by batch-size distribution is given by

$$a_{batch} = \sum_{k=1}^{3} b_k a_{simple,k} \qquad (B.3)$$

where

k = index of batch sizes of respondents

k = 1, custom and small-batch production

k = 2, large-batch production

k = 3, mass production

b_k = percentage of value added in the metalworking sector produced by industries dominated by batch-size class j. For example, $b_1 = 0.579 =$ percentage of value added in the metalworking sector produced by industries dominated by small-batch production.

$a_{simple,k}$ = simple average of substitution estimates of respondents who are in batch-size class k.

With respect to the distribution of employment by size of establishment and to the distribution of value added by mode of production, the characteristics of the establishments who responded to this part of the survey sample are noticeably different from the characteristics of the broader population of establishments throughout the metalworking industries.[3] These differences can be inferred from the data in Tables B-3, B-4, B-5. In the survey sample, over 62 percent of the establishments responding had 1000 or more production workers, and the remainder were distributed over the other three size classes. If one were to count the total number of production workers covered by plants in the survey, almost all of them would be employed in plants with 1000 or more workers.[4]

3. The characteristics of the full sample of 52 respondents will be discussed later. The conclusions stated here about the characteristics of robot users based on the subset of 24 respondents are consistent with the characteristics of robot users based on the full sample of 52 respondents.

4. A lower bound for the proportion of workers in the survey employed in plants with 1000 or more workers is 77 percent. This figure is derived by assuming that there are three establishments

According to the employment figures in Table B-4, as of 1977 fewer than 40 percent of all production workers in SIC 34–37 were employed in establishments with 1000 or more production workers. Thus, the proportion of production workers employed in large establishments (1000 or more production workers) in the sample survey is *over twice* the proportion for the metalworking industries as a whole.

Nine of the 24 survey respondents classified themselves as mass producers. On the assumption that each establishment in each batch-size region generated the same level of value added, the nine mass producers would have contributed 37.5 percent of the value added covered by the establishments in the sample. However, given the more realistic assumption that substantially more total value added is generated by the mass-production establishments than by the others, the nine mass producers would have accounted for substantially more than 37.5 percent of the value added by the establishments in the sample. According to the estimates of value added by mode of production across the entire metalworking sector made in Chapter 3 and summarized in Table B-5, approximately 25 percent of the value added is generated by industries dominated by mass production. Thus, the proportion of value added that is generated by mass-production establishments in the survey appears to be substantially larger than the proportion for the metalworking industries as a whole.

The preceding discussion of the characteristics of establishments in the sample vis-à-vis those in the universe of metalworking establishments has bearing on which technique to choose to average the substitution estimates. The simple average (Eq. B.1), which weights each response equally, would be the adequate if the characteristics of the sample closely resembled those of the population of interest. As noted above, this is clearly not the case. Therefore, the simple average is not an ideal way to average the substitution estimates. Because of the attributes of the sample, the simple average gives disproportionate weight to large establishments and to mass producers. Thus, a weighted average of responses should be used where the weights reflect the attributes of the universe of metalworking establishments. The question then becomes which of the two weighted averages should be used.

There are several reasons for arguing that the classification of respondents by the number of production workers in the establishment is less ambiguous than the classification by average batch size. Within a plant, the number of production workers is easier to count and more widely known than the average batch

---

with 99 workers, 4 establishments with 499 workers, 2 establishments with 999 workers, and 15 establishments with 1000 or more workers. Since some of the establishments in the largest size class have many more than 1000 production workers (for example, Lockheed and Caterpillar both had over 5000 at the time of the survey) and the establishments in the smaller classes do not all have the maximum number, it is clear that almost all of the production workers in the sample are employed in firms with 1000 or more workers.

Table B-5: Breakdown of Survey Respondents by Size of Establishment and Average Batch Size

	Number of Survey Respondents	Percentage of Survey Respondents
Size of Establishment (number of production workers)		
1–99	3	12.5
100–499	4	16.7
500–999	2	8.3
≥ 1000	15	62.5
Average Batch Size		
Small batch (1–1000)	9	37.5
Large batch (1001–10,000)	6	25.0
Mass (> 10,000)	9	37.5

size, so one would expect there to be fewer errors in the counts of production workers given by the respondents. Also, at a given point in time, an establishment is typically characterized by an average batch size *distribution* as opposed to a simple average batch size (see the footnotes to Table B-1). However, the number of production workers at a given point in time is a single number. The distribution of employees by size of establishment for the universe of metalworking industries given by the *Census of Manufactures* is a much more precise estimate than the distribution of value added by mode of production given in Chapter 3. As a result, the weights, s_j, in Equation B.2 are more precise then the weights, b_k, in Equation B.3. These factors favor the weighted average based on the distribution of employment by size of establishment over those based on the distribution by batch size.

With these acknowledgments, there are still well-founded reasons for arguing that the average based on the distribution by batch size is preferred to the average based on the distribution of employees by size of establishment. From a technological and engineering perspective, the volume of output (or average batch size) is often cited as the most important factor in determining the suitability of a particular type of technology for a particular type of job. This would suggest that batch size plays a more important role in determining the technical potential for substituting robots than does the number of workers in the establishment, since there is no clear correspondence between the size of the establishment and the mode of production. As for the problem of batch sizes being described by a distribution, one could argue that the output of most establishments falls within a range of one to two orders of magnitude. Thus, output can be characterized by a gross description of batch size (custom, batch, or mass). There is also an operational concern. Fifteen of the 24 respondents are concentrated in the largest establishment size. While the average estimate obtained for this one size class is reasonably reliable because of a relatively large sample size, the quality of the average estimates for the other size classes is necessarily questionable because of the very small sample sizes. The 24 respondents are more evenly distributed across the three batch sizes, so the quality of the average estimates within each batch-size range is more consistent.

In view of the above discussion, does the use of one average or the other make a large enough difference in estimating the number of displaced workers to make the choice worth worrying about? For the most part, no, since for most occupations the average estimates of potential substitution are very close to one another. However, for a few occupations, such as painters, riveters, and tool and die makers, the differences between the two weighted averages are not inconsequential. To facilitate computations and exposition, one type of average will be used. All calculations of the potential for substituting robots for workers will be based on the averages obtained by weighting the survey responses by the distribution of employment by size of establishment.

Appendix C
Estimated Potential Displacement of Production Workers in the Metalworking Industries, SIC 34–37

In this appendix, an estimate is made of the percentage and number of jobs that could be performed by Level I and Level II robots in the metalworking industries. All production worker occupations for SIC 34–37 listed in the 1980 *Occupational Employment* survey (Bureau of Labor Statistics, 1982b) are classified into nine groups:

Tool handlers
Metalcutting machine operators
Metalforming machine operators
Other machine operators
Assemblers
Inspectors
Material handlers and laborers
Miscellaneous craftworkers
Maintenance and transport workers

Within each of the nine groups of production workers, the estimates of the percentage of workers that could be displaced within each occupation are based directly on the survey, if survey data were available. (The survey estimates are given in Appendix B.) If survey responses were not collected for a particular occupation, then the potential for robot use in that occupation is estimated from the choice of a surrogate occupation. The average estimates of potential displacement for the group are derived from the total number who could be displaced and from the total employed within the group.

Table C-1 first shows summary of the estimates of the potential percentage of displacement within each major group of occupations, followed by a detailed breakdown of the specific occupations within each group and the estimates of potential for displacement within each occupation. Most of those who could be displaced in SIC 34–37 are metalcutting machine tool operators, other machine operators, tool handlers, metalforming machine operators, and assemblers. The nature of the jobs of maintenance and transport workers are such that they could not be performed by robots. In this discussion, the possibility that more maintenance workers would be required to support the robot applications is not considered. Also, only a small percentage of those workers classified as miscellaneous craftworkers could be displaced by robots. Given the procedure for estimating potential displacement and the employment figures for 1980, it is estimated that about 12 percent of the production workers could be displaced by Level I robots and 33 percent by Level II robots.

Table C-1: Summary and Occupational Breakdown of the Estimated Potential Displacement of Production Workers in the Metalworking Industries (SIC 34–37)

Occupation	Total Employment 1980	Potential Number of Workers Displaced by:		Potential Percentage of Workers Displaced by	
		Level I	Level II	Level I	Level II
Tool handlers	383,490	113,174	190,973	29.5	49.8
Metalcutting machine operators	1,073,580	153,422	437,661	14.3	40.8
Metalforming machine operators	337,841	110,553	207,195	32.7	61.3
Other machine operators	785,660	126,732	287,448	16.1	36.6
Assemblers	974,410	86,722	280,630	8.9	28.8
Inspectors	216,800	16,260	76,747	7.5	35.4
Material handlers and laborers	479,960	15,887	197,271	3.3	41.1
Miscellaneous craft workers	507,980	2,625	31,234	0.5	6.1
Maintenance and transport workers	362,250	0	0	0.0	0.0
Total	5,121,971	625,374	1,709,158	12.2	33.4
Tool Handlers					
Painters, production	75,250	32,734	50,267	43.5	66.8
Power screwdriver operators[a]	7,960	3,224	4,123	40.5	51.6
Sandblasters and shotblasters	6,380	2,271	2,271	35.6	35.6
Welders and flamecutters, total	293,900	74,945	134,312	25.5	45.7
All tool handlers	383,490	113,174	190,973	29.5	49.8

(continued on the following page)

Table C-1: Summary and Occupational Breakdown of the Estimated Potential Displacement of Production Workers in the Metalworking Industries (SIC 34–37) (*continued*)

Occupation	Total Employment 1980	Potential Number of Workers Displaced by:		Potential Percentage of Workers Displaced by	
		Level I	Level II	Level I	Level II
Metalcutting Machine Operators					
Die sinkers[b]	1,550	20	237	1.3	15.3
Drill press/boring machine operators	115,880	37,661	77,640	32.5	67.0
Filers, grinders, buffers, chippers	76,050	7,453	20,990	9.8	27.6
Gear-cutting, gear-grinding and/or gear-shaping machine operator[c]	13,840	2,519	7,958	18.2	57.5
Grinding/abrading machine operators, metal	105,610	19,221	60,726	18.2	57.5
Lathe machine operators, metal	140,800	25,626	70,963	18.2	50.4
Machine tool operators, combination[d]	152,400	27,737	76,810	18.2	50.4
Machine tool operators, toolroom[b]	34,320	446	5,251	1.3	15.3
Machine tool operators, NC	68,280	14,817	31,750	21.7	46.5
Machine tool setters[b]	53,300	693	8,155	1.3	15.3
Machinists[b]	96,930	1,260	14,830	1.3	15.3
Milling and planing machine operators	62,750	10,103	32,693	16.1	52.1
Patternmakers, metal[b]	5,980	78	915	1.3	15.3
Sawyers, metal[e]	12,480	4,056	8,362	32.5	67.0
Tool and die makers	133,210	1,732	20,381	1.3	15.3
All metalcutting machine operators	1,073,580	153,422	437,661	14.3	40.8
Metalforming Machine Operators					
Automatic spring coiling machine operators[f]	2,960	1,311	2,220	44.3	75.0
Bodymaker operators, tin can[f]	3,350	1,484	2,513	44.3	75.0
Die setters[g]	4,180	63	640	1.5	15.3
Forging press operators[g]	5,850	942	3,042	16.1	52.0
Forging/straightening roll operators[g]	1,740	280	905	16.1	52.0
Hammersmiths, open die[g]	2,710	436	1,409	16.1	52.0
Header operators[g]	3,820	615	1,986	16.1	52.0
Multislide machine operators[f]	2,310	1,023	1,733	44.3	75.0
Power brake/bending machine operators[f]	39,041	17,295	29,281	44.3	75.0
Punch press operators, metal	139,520	61,807	104,640	44.3	75.0
Punch press setters, metal[h]	18,800	282	2,876	1.5	15.3
Riveters, heavy	4,000	1,660	2,072	41.5	51.8
Riveters, light	8,280	3,436	4,289	41.5	51.8
Roll forming machine operators[f]	4,380	1,940	3,285	44.3	75.0
Shear and slitter operators, metal[f]	21,350	9,458	16,013	44.3	75.0
Sheer and slitter setters[h]	5,710	86	874	1.5	15.3
Sheet metalworkers	65,610	6,561	26,244	10.0	40.0
Wire drawers[f]	1,820	806	1,365	44.3	75.0
Wire weavers[f]	2,410	1.068	1,808	44.3	75.0
All metalforming machine operators	337,841	110,553	207,195	32.7	61.3

(*continued on the following page*)

Table C-1: Summary and Occupational Breakdown of the Estimated Potential Displacement of Production Workers in the Metalworking Industries (SIC 34–37) (continued)

Occupation	Total Employment 1980	Potential Number of Workers Displaced by:		Potential Percentage of Workers Displaced by	
		Level I	Level II	Level I	Level II
Other Machine Operators					
Balancing machine operators[i]	1,770	322	892	18.2	50.4
Coil finishers[j]	14,470	3,545	5,817	24.5	40.2
Coil tapers, hand or machine[j]	3,510	860	1,411	24.5	40.2
Coil winders	27,900	6,836	11,216	24.5	40.2
Compressor/injection molding machine operators[k]	20,140	8,922	15,105	44.3	75.0
Coremakers, machine[l]	1,560	20	239	1.3	15.3
Die casting machine operators and setters[k]	5,950	2,636	4,463	44.3	75.0
Dip platers, nonelectrolytic	8,100	4,512	6,456	55.7	79.7
Electroplaters	29,240	5,292	12,544	18.1	42.9
Encapsulaters[m]	5,870	3,270	4,678	55.7	79.7
Etchers and/or engravers	5,850	1,580	3,101	27.0	53.0
Furnace operators, cupola tenders	4,060	118	597	2.9	14.7
Heat treaters, annealers, temperers	15,230	3,457	9,306	22.7	61.1
Heaters, metal[n]	3,850	874	2,352	22.7	61.1
Impregnators, electronic[o]	2,010	364	862	18.1	42.9
Industrial truck operators	88,980	0	0	0.0	0.0
Loading machine operators[o]	810	359	608	44.3	75.0
Mixers and/or blenders, chemicals and chemical products[p]	2,460	148	246	6.0	10.0
Metal molders, machine[q]	6,170	531	963	8.6	15.6
Oilers	6,160	0	0	0.0	0.0
Plater helpers[r]	22,170	0	11,085	0.0	50.0
Pourers, metal	2,970	324	636	10.9	21.4
Press operators and/or plate printers[s]	3,790	49	580	1.3	15.3
Production packagers	54,760	5,914	14,511	10.8	26.5
Sand cutters, mixers and/or slingers[t]	630	115	362	18.2	57.5
Screen/stencil printers/setters[u]	3,470	937	1,839	27.0	53.0
Sewing machine operators, regular equipment	2,560	0	0	0.0	0.0
Sewing machine operators, special equipment[v]	740	66	213	8.9	28.8
Stationary boiler firers, nonmetal[w]	1,620	47	238	2.9	14.7
Testers	61,560	2,955	6,956	4.8	11.3
Wirers, electronic	28,760	2,013	7,938	7.0	27.6
Woodworking machine operators[i]	4,170	759	2,102	18.2	50.4
All other operatives and semiskilled workers[x]	344,370	69,907	160,132	20.3	46.5
All other machine operators	785,660	126,732	287,448	16.1	36.6
Assemblers					
All assemblers	974,410	86,722	280,630	8.9	28.8
Inspectors					
All inspectors	216,800	16,260	76,747	7.5	35.4

(continued on the following page)

5

5555555554544444

445

Table C-1: Summary and Occupational Breakdown of the Estimated Potential Displacement of Production Workers in the Metalworking Industries (SIC 34–37) (*continued*)

Occupation	Total Employment 1980	Potential Number of Workers Displaced by: Level I	Level II	Potential Percentage of Workers Displaced by Level I	Level II
Material Handlers and Laborers					
Conveyor operators and tenders	11,380	1,696	4,768	14.9	41.9
Furnace operators and heater helpers[y]	1,610	0	805	0.0	50.0
Helpers, trades	63,840	0	31,920	0.0	50.0
Order fillers	27,020	1,864	8,565	6.9	31.7
Riggers	8,230	0	0	0.0	0.0
Shakeout workers, foundry[z]	1,610	240	675	14.9	41.9
All other laborers[aa]	366,270	12,087	150,537	3.3	41.1
All material handlers	479,960	15,887	197,271	3.3	41.1
Miscellaneous Craft Workers					
Blacksmiths[bb]	950	12	145	1.3	15.3
Blue-collar work supervisors	258,630	0	0	0.0	0.0
Boilermakers[bb]	6,050	79	926	1.3	15.3
Coremakers (hand, bench, floor)[bb]	2,330	30	356	1.3	15.3
Crane, derrick, and hoist operators	19,990	0	0	0.0	0.0
Fabricators, plastics[bb]	5,350	70	819	1.3	15.3
Fitters, structural metal[bb]	14,440	188	2,209	1.3	15.3
Glaziers[bb]	1,630	21	249	1.3	15.3
Instrument repairers[bb]	4,130	54	632	1.3	15.3
Laminators, preforms[bb]	10,930	142	1,672	1.3	15.3
Layout markers, metal[bb,cc]	16,220	211	2,482	1.3	15.3
Layout markers, wood[bb]	180	2	28	1.3	15.3
Metal fabricators[bb]	11,530	150	1,764	1.3	15.3
Millwrights[bb]	31,570	410	4,830	1.3	15.3
Molders, metal (bench, floor)[bb]	3,980	52	609	1.3	15.3
Patternmakers, wood[bb]	5,220	68	799	1.3	15.3
Refractory material repairers[bb]	690	9	106	1.3	15.3
Shipfitters[bb]	15,420	200	2,359	1.3	15.3
Shipriggers[bb]	2,570	33	393	1.3	15.3
Shipwrights[bb]	2,830	37	433	1.3	15.3
Structural steel workers[bb]	4,040	53	618	1.3	15.3
All other skilled craft and kindred workers[dd]	89,300	804	9,805	0.9	11.0
All miscellaneous craft workers	507,980	2,625	31,234	0.5	6.1
Maintenance and Transport Workers					
Carpenters	20,510	0	0	0.0	0.0
Delivery and route workers	5,980	0	0	0.0	0.0
Drivers, assembly line	760	0	0	0.0	0.0
Electricians	58,380	0	0	0.0	0.0
Maintenance repairers, general utility	46,080	0	0	0.0	0.0
Mechanics and repairers	153,290	0	0	0.0	0.0
Painters, maintenance	7,950	0	0	0.0	0.0
Plumbers and/or pipefitters	30,200	0	0	0.0	0.0
Stationary engineers	4,480	0	0	0.0	0.0
Truck drivers	34,620	0	0	0.0	0.0
All maintenance and transport workers	362,250	0	0	0.0	0.0

(*continued on the following page*)

Table C-1: Summary and Occupational Breakdown of the Estimated Potential Displacement of Production Workers in the Metalworking Industries (SIC 34–37) (*continued*)

[a]Estimate based on survey response for riveters.
[b]Estimate based on survey response for tool and die makers.
[c]Estimate based on survey response for grinding/abrading machine operators.
[d]Machine tool operators who alternatively operate lathes, boring machines, grinding machines, and milling machines. Estimate based on survey response for lathe machine operators.
[e]Estimate based on survey response for drilling machine operators.
[f]Estimate based on survey response for punch press operators.
[g]Estimate based on survey response for milling machine operators.
[h]Estimate based on survey response for tool and die makers.
[i]Estimate based on survey response for lathe operators.
[j]Estimate based on survey response for coil winders.
[k]Estimate based on survey response for punch press operators.
[l]Estimate based on survey response for tool and die makers.
[m]Estimate based on survey response for dip platers.
[n]Estimate based on survey response for heat treaters.
[o]Estimate based on survey response for electroplaters.
[p]Estimate based on survey response for mixers.
[q]Estimate based on survey response for casters.
[r]Estimate based on survey response for helpers, trades.
[s]Assumes this refers to printing press operators. Estimate based on survey response for tool and die makers.
[t]Estimate based on survey response for grinding/abrading machine operators.
[u]Estimate based on survey response for etchers/engravers.
[v]Estimate based on survey response for assemblers.
[w]Estimate based on survey response for furnace operators/cupola tenders.
[x]Estimate based on average percent displacement for the specified categories of other machine operators.
[y]Estimate based on survey response for helpers, trades.
[z]Estimate based on survey response for conveyor operators and tenders.
[aa]Estimate based on average percent displacement for specified laborers.
[bb]Estimate based on survey response for tool and die makers.
[cc]Layout markers are only used with manually controlled machine tools. All workers would be eliminated if all machines were numerically controlled.
[dd]Excluding etchers and/or engravers, which are listed under other machine operators.

Appendix D
An Estimated Percentage of Metalcutting Machine Tools That Could Be Operated by Robots

In this appendix, a procedure is described for estimating lower and upper bounds on the percentage of metalcutting machine tools that could be operated by a robot. This estimate makes no use of the survey data described in the previous appendix and is made in order that it can be compared with the survey estimates of the percentage of metalcutting machine operators that could be replaced by robots.

The *American Machinist* (1978) estimates the number of machines in use for several major categories of equipment for each of the metalworking industries. Metalcutting tools are divided into 19 major categories, and most of these major categories are further subdivided. All together, 97 types of metalcutting machine tools are enumerated in the 12th *American Machinist* Inventory (*American Machinist, 1978*), which was conducted from 1976 to 1978. The counts of the total number of units in each category were derived from sample survey responses from 20 percent of all metalworking establishments with 20 or more employees. Over 7600 establishments responded to the survey.

I assembled information on the function and operation of almost all of the 97 types of metalcutting machines from reference books on metalworking technologies (Roberts and Lapidge, 1977; American Society for Metals, 1967) and from interviews with manufacturing experts.[1] Decisions on whether or not a particular type of machine would be a candidate for robotic operation were made in the following fashion: Four different categories of machine types were defined. (Their definitions are presented in detail below.) I decided whether or

1. Robert McDivett, former director of manufacturing for Westinghouse Power Systems, was the primary source of "practical" information on machine tool uses. Jim Brecker, of the Westinghouse Productivity and Quality Center, and Charles Carter, technical director of Cincinnati Milacron, also provided information.

not robots would, in general, be used to operate machines within a given category, assuming the robot would be retrofitted into an existing facility. This decision was subjective, but it was based on practical manufacturing considerations that include how the current stock of machines are typically used and on previous experience where robots have proven to be economically viable alternatives to machine operators. I assigned each of the separately enumerated machine types within the *American Machinist's* 19 major categories of metalcutting machines to at least one of my four categories. In some cases, a type of machine may belong to two, possibly even three, categories (see Table D-4). In such cases, all the applicable categories are listed. For example, a single-spindle automatic screw machine may be used either as a Category 2 machine (very high volume production; operates automatically without operator) or as a Category 4 machine (medium- and large-batch production; requires operator to load and unload each piece after completion of cycle), depending on the particular type of machine, the way it is used in a particular factory, and the type of part being made. I made no attempt to estimate how all of the single-spindle automatic screw machines throughout all of the metalworking industries are distributed between Category 2 and Category 4. Therefore, a minimum estimate of the number of Category 2 machines does not include the number of single-spindle automatic screw machines (since they might all be used as Category 4 machines), however, a maximum estimate of the number of Category 2 machines does. Following through with this procedure, I estimated the minimum and maximum percentage of metalcutting machines within each of the four categories (Table D-1). With these estimates and with the rules for deciding on which categories of machines would be candidates for robot operation, I calculated the lower and upper bounds on the percentage of metalcutting machines that are candidates for robot operation.

The following are definitions of the four categories of metalcutting machines:

Category 1: machines that are designed to operate most efficiently for "piece" and small-lot production of parts that are of small to moderate size and weight (they can be loaded and unloaded by a human operator).[2] Within a typical day, the time spent setting up these types of machines (e.g., most engine and tool room lathes) exceeds the time they are in operation cutting metal.

These types of machines would seldom be suitable for robot applications given the current organization of the flow of materials and the scheduling of

---

2. "Small lot" is not defined because it depends on the type of part being manufactured. For example, if a complicated shape requiring several different tools were being made, a small lot may consist of two or three parts. If a simple shape requiring one type of tool (therefore eliminating the need for tool change) were being made, a small lot may consist of 100–200 parts.

Table D-1: Low and High Estimates of the Percentage Distribution of Metalcutting Machine Tools by Category

	Low Estimate	High Estimate
Category I (machines designed for low-volume production)	39.4	68.2
Category II (machines designed for fully automatic operation)	12.0	19.9
Category III (machines designed for very large and/or heavy workpieces)	1.1	1.7
Category IV (machines designed for medium- to large-batch production)	9.4	46.7

parts and tools in virtually all factories. Even though the tasks required to load and unload a particular part might be such that it could, in principal, be performed by a Level I robot, in almost all cases it would take longer to program the robot to do the job than for a human operator to do it once. Since setup and retooling occur so frequently, an operator would be required to attend the machine nearly full time anyway to change machine settings and tools. Also, because of the need for frequent setups and short production runs, the use of special-purpose tools and fixtures and the setting of mechanical stops and controls are kept to a minimum, which leaves many of the machining functions, such as control of feeds and speeds, under the control of the operator. Neither a Level I nor Level II robot would be able to to operate the machine unless the manual controls of machining cycles, feeds, and speeds were replaced by numerical control.

Category 2: machines that are designed to operate under fully automatic control, without the assistance of a human operator once they are set up. They include "special purpose" machines designed for very high volume operations where the loading, unloading, and cutting operations are already highly automated (e.g., transfer machines and most types of automatic screw machines and cut-off saws). "Speciality" machines, such as gear makers and electrochemical machines, which run under automatic cycles for very long periods of times and which do not require constant monitoring by an operator, are also included in this category.

These types of machines would seldom be suitable for robot applications, since the machine typically runs unattended once it is set up. In practice, one operator is usually responsible for setting up and monitoring several of these machines.

Category 3: machines designed to handle parts that are very large and/or heavy (e.g., vertical and horizontal boring mills). The loading and unloading of parts on these machines would not be suitable for robot applications. These parts are often so large that a large crane is required to move them. Loading and unloading is complicated by the fact that, prior to precision machining, the workpiece is usually not in strict conformance with the specifications. A decision is often made during loading as to how to position the piece on the machine in order to compensate for imprecisions. Even if very heavy-duty robots were available, part loading could not be robotized because of the very complicated and variable sensing and decision-making requirements. Once the part is loaded and properly positioned on the machine, the cutting time is typically very long (often several hours). An operator is required to monitor the operation constantly for signs of problems with the part or with the tool, and to change tools when necessary. A robot with the ability to monitor the cutting process and to make the necessary adaptations to maintain the proper conditions would be able to operate the machine during the cutting cycle. However, this would probably require sophistication beyond that of a rudimentary Level II robot.

Category 4: the remaining types of machines. By omission, these are the machines that are *not* designed for:
  small-volume production (Category 1),
  fully automatic operation (Category 2), and
  processing very large and/or very heavy workpieces (Category 3).
  Category 4 includes machines that produce parts of moderate sizes and shapes in medium to large batches and requires regular loading and unloading as part of these operations cycle. Examples of these types of machines include most numerically controlled machines and automatic between-centers chucking machines.

Most of these types of machines are well suited to be operated by a robot, since the machining cycle is carried out automatically once the part is placed in the machine. (The machine need not be numerically controlled to have an automatic cycle. It can by controlled by other means.) In most cases, an operator would be required to set up the machine. After that, the loading and unloading of the part could be be carried out with a robot. Some additional tasks, such as "off-line" inspection, could be performed by the robot as well.

Level I robots could be used to operate Category 4 machines if all aspects of the tasks of acquiring parts and placing them in holding fixtures were completely determined and could be prespecified in advance of performing the operation.

Level II robots would be required to operate Category 4 machines if it were not possible or economically feasible to completely determine and prespecify all aspects of the tasks of acquiring parts and placing them in holding fixtures once the machine has been set up. This would make it necessary for the robot to sense the position and/or orientation of the part and to dynamically adjust its instructions to accomplish the task.

Rules of thumb about the applicability of robots to each category of machines are summarized as follows:

Category 1   Neither Level I nor Level II robots would be used to operate these types of machines, because robot programming costs and special tooling costs would outweigh the benefits of replacing the operator. In addition, there is the difficulty of operating manually controlled machines.

Category 2   Neither Level I nor Level II robots would be used to operate these types of machines since their operation is already fully automated, requiring minimal operator assistance, except for setup.

Category 3   Neither Level I or Level II robots designed for parts acquisition and parts placement would be used to operate these machines.

Category 4   Level I robots could be used to operate these types of machines if all aspects of acquiring the part and positioning it in the workholding device were completely determined and could be prespecified. If these aspects could not be completely determined and prespecified, Level II robots could be used to operate these types of machines. Rudimentary Level II robots *could not* operate these types of machines if they were required to do complicated setup procedures.

There are clearly exceptions to these rules. For example, even Category 1 machines could be operated by Level I robots if parts were designed so that one type of gripper could be used, if manufacturing requirements were planned to minimize tool changes, and if parts were scheduled so as to maximize the number of distinct parts that could be processed on the machine for a given setup. However, this requires a far more extensive reorganization of production (which may include the redesign of the product) than is being considered here. For this reason, it is conservatively assumed that Category 1 machines would not be retrofitted with robots. In addition, there might be ways in which robots could be used to aid in the operation of Category 2 machines (e.g., inspecting and/or palletizing parts when they come off the machine). However, none of these applications are being considered here.

While it is assumed that Category 4 machines could be operated by robots, there is no unambiguous way of designating what fraction of the Category 4 machines could be operated by Level I robots and what fraction would require Level II robots. One could assume that all Category 4 machines could, in principal, be operated by a Level I robot if the feeding of parts to the robot were controlled and if the work-holding devices were appropriately designed. The question arises as to how to decide when it would be worth the cost and effort to rationalize the workplace so that the machine could be operated by a Level I robot. Where it would not be worth the cost and effort to make these changes, a Level II robot would be required. One plausible scheme is to assume that if production runs were always in medium-sized to large batches, the cost of making the necessary adjustments could be justified, and a Level I robot could be used. If production runs were sometimes for small batches, as well as for medium and large batches, then it would be too time consuming and expensive to always make the necessary adjustments for an insensate robot. This would suggest that Level I robots could be used to operate machines exclusively assigned to Category 4 and machines jointly assigned to Categories 4 and 2. However, Level I robots could not be used to operate machines jointly assigned to Categories 4 and 1 or Categories 4 and 3.

It is assumed that Level II robots could be used to operate any machine that could be operated by a Level I robot. Since Level II robots have some capability to sense the work environment and to acquire and/or orient parts even though all aspects of the task are not precisely prespecified, it is assumed that they could be used to operate a machine if production runs were sometimes for small batches in addition to medium and large batches. This suggests that, in addition to the machines that could be operated by Level I robots, Level II robots could operate machines that are jointly assigned to Categories 4 and 1 and Categories 4 and 3.

Machines designated as belonging only to Category 4 constitute 9.4 percent of the 12th *American Machinist* Inventory. Types of machines which are often used as Category 4 machines, but which might also be used in another category are not included in this 9.4 percent. For example, turret lathes can be used as either a Category 1 or Category 4 machine, automatic screw machines can be used as either a Category 2 or Category 4 machine, and planers can be used as either a Category 3 or Category 4 machine. These other types of machines which are jointly assigned to Category 4 and to another category constitute 37.3 percent of the metalcutting machines. Thus, the analysis yields that at the very most, 46.7 percent of the machine tools are Category 4 machines. This is used as an upper estimate of the percentage of metalcutting machine tools which could be operated by Level II robots. Of the 37.3 percent of machines that are jointly assigned to Category 4 and another category, 6.3 percent are jointly assigned to Category 4 and Category 2. Since these machines are used for medium to very large batch production, it is assumed that they could also be operated by Level I robots. Together, machines assigned exclusively to Category 4 and jointly to Categories 4 and 2 compose 15.7 percent of the metalcutting machines. This is used as an upper estimate of the percentage of machines that could be operated by Level I robots. The estimated percentage of metalcutting machines that could be operated by Level I and Level II robots is shown in Table D-2.

This estimated percentage of metalcutting machine tools which could be operated by a robot is now compared with the estimated percentage of metalcutting machine operators who could be displaced by a robot, which is derived

Table D-2: Estimated Percentage of Metalcutting Machine Tools That Could Be Operated by Level I and Level II Robots

Machine Types Assigned to:	Percentage of All Metalcutting Machines in 12th *American Machinist* Inventory
Category 4 only	9.4
Categories 4 and 2	6.3
Subtotal	15.7
Categories 4 and 1, 4 and 2 and 4 and 3	37.3
Total, Category 4 (exclusively and jointly)	46.7

Machines which could be operated by a Level I robot = 9.4–15.7%
Machines which could be operated by a Level II robot = 46.7%

from the 1981 CMU robotics survey. The 15 occupational titles in the Bureau of Labor Statistics (1982b) *Occupational Employment* survey which cover the workers who operate metalcutting machine tools are listed in Table D-3. The survey results indicate that 153,400 of their jobs could be performed by Level I robots and 437,600 by Level II robots. This corresponds to displacement rates of 14.3 percent and 40.8 percent, respectively.

The estimated percentage of metalcutting machine tools that could be oper-

Table D-3: Potential Displacement of Metalcutting Machine Operators by Robots in SIC 34–37

	Total Employment 1980	Potential Number of Workers Displaced by:		Potential Percentage of Workers Displaced by:	
		Level I	Level II	Level I	Level II
Die sinkers[a]	1,550	20	237	1.3	15.3
Drill press/boring machine operators	115,880	37,661	77,640	32.5	67.0
Filers, grinders, buffers, chippers	76,050	7,453	20,990	9.8	27.6
Gear-cutting, gear-grinding, and/or gear-shaping machine operators[b]	13,840	2,519	7,958	18.2	57.5
Grinding/abrading machine operators, metal	105,810	19,221	60,726	10.2	57.5
Lathe machine operators, metal	140,800	25,826	70,963	18.2	50.4
Machine tool operators, combination[c]	152,400	27,737	76,810	18.2	50.4
Machine tool operators, toolroom[a]	34,320	446	5,251	1.3	15.3
Machine tool operators, NC	68,280	14,817	31,750	21.7	46.5
Machine tool setters[a]	53,300	693	8,155	1.3	15.3
Machinists[a]	96,930	1,280	14,830	1.3	15.3
Milling and planing machine operators	62,750	10,103	32,693	18.1	52.1
Patternmakers, metal[a]	5,980	78	915	1.3	15.3
Sawyers, metal[d]	21,480	4,056	8,362	32.5	67.0
Tool and die makers	133,210	1,732	20,381	1.3	15.3
All metalcutting machine operators	1,073,580	153,422	437,661	14.3	40.8

[a]Estimate based on survey response for tool and die makers.

[b]Estimate based on survey response for grinding/abrading machine operators.

[c]Machine tool operators who alternatively operate lathes, boring machines, grinding machines, and milling machines. Estimate based on survey response for lathe machine operators.

[d]Estimate based on survey response for drilling machine operators.

ated by Level I robots closely brackets the estimated percentage of metalcutting machine tool operators that could be displaced by Level I robots. The estimated percentage of metalcutting machine tools that could be operated by Level II robots is moderately larger than the survey-based estimate of the percentage of machine tool operators that could be displaced by Level II robots. From this comparison, there is no reason to suspect that the survey estimates are unreasonable. Indeed, one might say that these survey estimates are good indicators of the potential for the use of Level I and Level II robots in machine operations. I have made no further attempts in this study to compare survey estimates for other groups of production workers with independently derived estimates of the potential for substitution. However, assuming that the survey estimates for metalcutting workers are credible, there is no reason to expect that the other estimates are not.

Table D-4: Category Assignment of the Metalcutting Machine Tools in the *12th American Machinist* Inventory[a]

Types of Machines	Category	
	Not NC Controlled	NC Controlled
Turning machines		
Bench	1	4
Engine and tool room, < 8 inch swing	1	4
Engine and tool room, 9–16 inch swing	1	4
Engine and tool room, 17–23 inch swing	1	4
Engine and tool room, ≥ 24 inch swing	1	4
Tracer lathe	1	4
Turret lathe, ram type	1, 4	4
Turret lathe, saddle type	1, 4	4
Automatic chucking, vertical and horizontal, single spindle	2, 4	2, 4
Automatic chucking, vertical and horizontal, multispindle	2, 4	2, 4
Automatic between-centers chucking	4	4
Automatic bar (screw) machine, single spindle	2, 4	2, 4
Automatic bar machine, multispindle	2, 4	2, 4
Vertical turning and boring mills (VTL, VBM)	3	3
Other (including forin, axle, spin, shell)	1, 4	4
Boring machines		
Horizonal bore, drill, mill (bar machine), table and planer type	1	4
Horizontal bore, drill, mill (bar machine), floor type	1	4
Precision, horizontal and vertical	1	4
Jig bore, horizontal and vertical	1	4
Other (not boring lathes)	1	4
Drilling machines		
Sensitive (hand feed), bench	1	4
Sensitive (hand feed), floor and pedestal	1	4
Upright, single spindle	1, 4	4
Upright, gang	4	4
Upright, turret, not NC	1, 4	—
Radial	1	4
Multispindle cluster (adjustable and fixed center)	2, 4	2, 4
Deep hole (including gun drill)	1, 2, 4	3, 4
Other (not unit head and way)	1, 4	4

(continued on the following page)

Table D-4: Category Assignment of the Metalcutting Machine Tools in the *12th American Machinist* Inventory[a] (*continued*)

Types of Machines	Category Not NC Controlled	NC Controlled
Milling machines		
Bench type (hand or power feed)	1	4
Hand	1	4
Vertical ram type (swivel head and turret)	1	4
General purpose, knee or bed:		
horizontal (pin, universal and ram)	1, 4	4
General purpose, knee or bed: vertical	1, 4	4
Manufacturing, knee or bed	1, 4	4
Planer type	1, 3	3, 4
Profiling and duplicating (including die, skin, spar)	4	4
Thread millers	2, 4	2, 4
Others (including spline, router, engraving)	1	4
Tapping machines	4	4
Threading machines	2, 4	2, 4
Multifunction NC machines (machining centers)		
Drill-mill-bore, manual tool changer,		
vertical and horizontal	4	4
Drill-mill-bore, indexing turret	4	4
Drill-mill-bore, auto tool changer, vertical	4	4
Drill-mill-bore, auto tool changer, horizontal	4	4
Special way type and transfer machines		
Single-station (several operations on one part)	2	2
Multistation, rotary transfer	2	2
Multistation, in-line transfer	2	2
Broaching machines		
Internal	4	4
Surface and other	4	4
Planing machines		
Double column	1, 3	3, 4
Open side and other	1, 3	3, 4
Shaping machines (not gear)		
Horizontal	1	4
Vertical, (slotters and keyseaters)	1	4
Cutoff and sawing machines		
Hacksaw	2	2
Circular saw (cold)	2	2
Abrasive wheel	2	2
Bandsaw	2	2
Contour sawing and filling	1	4
Other (including friction)	2	2
Grinding machines		
External, plain center type	1, 4	4
External, universal center type	1, 4	4
External, centerless (including shoe type)	4	4
External, chucking	1, 4	4
Internal, chucking (centerless shoe type)	1, 4	4
Surface, rotary table: vertical and horizontal	1, 4	4
Surface, reciprocating: horizontal manual	1	4
Surface, reciprocating: vertical and horizontal power	1	4
Disk grinders (not hand-held)	1	4

(*continued on the following page*)

Table D-4: Category Assignment of the Metalcutting Machine Tools in the *12th American Machinist* Inventory[a] (*continued*)

Types of Machines	Category	
	Not NC Controlled	NC Controlled
Grinding machines (*continued*)		
Abrasive belt (excluding polishing)	1, 4	4
Contour (profile)	4	4
Thread grinders	4	4
Tool and cutter	1	4
Bench, floor and snag	1, 4	4
Other (including jig)	1	4
Honing machines		
Internal (including combination bore-hone)	1, 4	4
External	1, 4	4
Lapping machines		
Flat surface	1	1
Cylindrical	1	1
Other (including combination hone-lap)	1	1
Polishing and buffing machines		
Polishing stands (bench and floor)	1	
Abrasive-belt, disk, drum (not grind)	4	
Other (including spindle lathes and multistation type)	2	2
Gear cutting and finishing machines		
Gear hobbers	2	2
Gear shapers	2	2
Bevel-gear cutters (including planer type)	2	2
Gear-tooth finish (grind, lab, shave, etc.)	2	2
Other gear cutting and finishing	2	2
Electrical machining units		
Electrical-discharge machines (EDM)	2	2
Electro-chemical machines (ECM)	2	2
Electrolytic grinders (ECG or ELG)	2	2

[a]Automatic assembly machines and "other" metalcutting machines are not included.

Appendix E

An Analysis of the Variation in Unit Cost across Metalworking Industries

The variation in unit cost across industries is examined in detail in this appendix. I have used a multiple linear regression model as the means of measuring how much of the variation in unit cost can be explained by the level of output and the nature of the processing activity, as well as by wage rates, the number of establishments, and the coverage of material inputs. The general form of the model is

$$\ln(u) = \beta_0 + \beta_1\ln(f_1) + \beta_2\ln(f_2) + \ldots + \beta_n\ln(f_n) + \epsilon \qquad (E.1)$$

where

$E[\ln(u)]$ = expected value of $\ln(u)$

u = unit cost

= value added/pounds of metal processed

$f_1, \ldots, f_n$ = explanatory factors, which are assumed to be known constants, with no exact linear relationship among themselves.

$$\beta_i = \frac{d \ln(u)}{d\ln(f_i)} \qquad (E.2)$$

= estimated elasticity of unit cost with respect to factor f_i

ϵ = error terms are assumed to be independent, normally distributed $N(0,\sigma)$

The parameters of the model $(\beta_1, \ldots, \beta_n)$ are measures of the "effect" associated with the factor alleged to be measured by each of the explanatory variables. If each explanatory variable were to increase by 1 percent, its relative influence on unit cost, with the level of all other variables held constant would be measured by the magnitude of its parameter. The multiple linear regression model only measures correlations between explanatory variables and the independent variable, and does not prove causal relations. Nonetheless, this framework still provides a well-established way of testing whether or not unit cost is related to each of the various factors and of estimating the magnitude of the effects of each factor.

Unit cost is inversely related to the level of output, as was seen in the comparison of unit cost and output across industries in Chapter 3. The regression results in Chapter 3 showed that the estimate of the elasticity of value added per pound of metal with respect to pounds of metal processed per establishment is negative and highly significant, supporting this hypothesis. However, the measure of pounds of metal processed per establishment accounts for only 30 percent of the variation in unit cost, which suggests that the level of output within an industry is only one of several factors which affect its level of unit cost. Considering the types of data used to construct surrogate measures of unit cost and output, interindustry differences in unit cost should also be related to interindustry differences in the nature of the processing requirements, in wage rates, and in the proportion of total material cost covered by purchases of basic and processed metals. There are also differences in the number of establishments across industries, but these differences are compensated for by normalizing pounds of metal processed within an industry by the number of establishments. Differences in the nature of the processing requirements cannot be directly measured given the available industry data, so the influence of this factor cannot be measured precisely. The influence of interindustry differences in the proportion of total material cost covered by purchases of basic and processed metals and by wage rates can easily be accounted for, since there are straightforward measures of the coverage of materials and of the hourly wage rate for each industry.

E.1. Explanatory Variables in the Regression Model

The rationale behind using pounds of metal processed as a surrogate measure of the scale of production and of modifying the measure to pounds of metal processed per number of establishments has already been discussed in Chapter 3. The explanatory variables used to represent the other factors which are believed to explain part of the interindustry differences in unit cost are discussed below. The bulk of the discussion focuses on proposed measures to indicate differences in the nature of the processing activities across industries.

The metal removal process is used to illustrate how differences in product attributes influence cost, with the level of output held constant. According to the *Machining Data Handbook* (Machinability Data Center, 1980: 21–23), in addition to the quantity of output produced, the cost of machining a metal product also depends on:

the material used

the size, rigidity, and geometry of the part

the accuracy and the surface roughness requirements

The material used affects cost because the amount of time it takes to remove a given amount of material depends on its metallurgical and "hardness" properties. "Softer" materials can be processed quickly and hence less expensively, because they can be cut at higher speeds, and they cause less wear on tools. Conversely, "harder" materials take longer to process and are consequently more expensive, because it takes longer to remove a given amount of material, and cutting tools must be replaced more frequently. For example, it may cost only half as much to machine a part made of aluminum alloy than a part made of plain carbon steel. Machining a part made of stainless steel or low alloy steel may be twice as expensive as machining a part of plain carbon steel. A part made of ultrahigh-strength steel may be more than five times as expensive to machine as a plain carbon steel part. In general, all types of specialty steels (high-strength steel, stainless steel, and alloy steel) are "harder" than plain carbon steel and more expensive to machine (Machinability Data Center, 1980: 21–28).

Very large and heavy parts require special cranes and blocking equipment to be moved and a long time to be positioned precisely on machines. Consequently, they are expensive to produce, especially if they require a very smooth surface finish or very close tolerances. Materials such as brass, which are less rigid and "slide" around when they are being machined, take longer to set up because they must be secured more carefully. Part geometry directly affects production cost, because the geometry of the finished product determines the amount of processing required to change the material from its original shape (bar, sheet, plate, etc.) to its final shape. The more severe the geometric change made to the material, the more processing time (and cost) is required. Also it sometimes takes more preparation to set up machines to produce more complicated geometric shapes. For example, if a part requires contouring (which requires that the cutting tool move simultaneously on the x and y coordinates), it is necessary either to prepare an NC tape or to make a template, whereas a simple squaring cut can be carried out without special preparation. Also, if a very irregular three-dimensional shape is required, such as for a die, a wooden pattern might have to be prepared.

Machining costs also increase as the surface roughness requirements become more stringent, because it takes more time and hence cost to achieve a smoother surface. For example, a "roughing" cut may have to be followed by an additional "semi-finishing" cut, and even by a third "finishing" cut to achieve the desired surface smoothness. For very smooth finishes, an additional operation, such as precision grinding, may be required. Achieving a very smooth surface finish (down to 0.000032 inches) from a very rough finish (0.002 inches) may increase machining cost by over 200 percent (Machinability Data Center, 1980: Sec. 21, 10). Increasing the accuracy of dimensional tolerances also increases cost for similar reasons. More than one cut may be required and/or the part may have to be transferred to more precise machines.

Clearly, for a given volume of output, a product with a simple shape made of plain carbon steel with a coarse surface roughness and a loose dimensional tolerance would be much less expensive than a product with a complicated shape made of ultrahigh-strength steel with a very fine surface finish and tight dimensional tolerances. This suggests that much of the variation in the relationships between unit cost and output per establishment across industries could be related to differences in the nature of processing requirements. Unfortunately, there are no conveniently available direct measures that can be constructed from industrywide data to indicate the relative complexity of materials processing across industries.

I have constructed surrogate measures of the relative complexity of processing requirements for an industry, because these differences exert such a large influence on unit cost. In the SIC data set of metalworking industries used in this analysis, the quantities and costs of the various metal inputs used by each industry are given, but there is no information on the size, geometry, accuracy, or surface roughness requirements of the products produced. Only the available information on the material inputs can be used to construct surrogate measures of the complexity of an industry's material processing.[1]

E.1.1. THE UNIT COST OF BASIC METAL INPUTS

The number of pounds of metal processed within an industry is the sum of two components: (1) pounds of *basic* metals and (2) pounds of *processed* metals. Basic metals include bars, plates, sheets, tubes, and other standard shapes made of steel, aluminum, and brass, as well as semifinished castings and forgings. Processed metal inputs are the finished products of other metalworking industries (such as screws, bearings, motors, control units, etc.) It is argued that the unit cost (dollars per pound) of the basic metal inputs purchased is an indicator of the relative complexity of the metal-shaping operations carried out within each industry. The foundation for this argument is that for ferrous metals there appears to be a strong correlation between the unit cost of the metal and the difficulty of machining it, and that in most industries ferrous metals account for almost all of the basic metal inputs. The major categories of ferrous metals used in SIC 34–37, and the average unit cost (dollars per pound of metal purchased)

---

1. For a few industries, it is possible to construct a measure of the average unit weight of the output by dividing the total pounds of metal processed by the total number of units shipped. This would serve as a good proxy of average product size. However, data on units shipped for all major classes of products are not given for many industries. Where the number of units shipped is given, it is for very detailed product classes, making it necessary to go through painstaking aggregation to get total units shipped for a major product class. At the three digit SIC level, information on the composition of the machine tools used in each industry group is available from the 12th American Machinist Inventory (Appendix D). An index of the relative degree of surface roughness required across industries could possibly be constructed from the ratio of the number of grinding, honing, and lapping machines to total machine tools.

Table E-1: Average Unit Cost of Ferrous Metal Inputs Used in SIC 34–37

Ferrous Metal Input	Unit Cost (in dollars per pound)			
	34	35	36	37
Stainless steels	0.97	1.03	0.92	0.73
Steel castings	0.59	0.96	1.00	0.51
Iron and steel forgings	0.52	0.64	0.57	0.44
Iron castings	0.24	0.49	0.34	0.33
Alloy steels	0.26	0.42	0.32	0.22
Carbon steels	0.19	0.21	0.19	0.18
Pig iron	0.09	0.09	—	0.09
Iron/steel scrap	0.06	0.04	—	0.04
Ferrous metal proportion of total basic metals[a]	93.0%	93.7%	94.8%	94.3%

Source: Derived from Tables E-3–E-6.

[a]Ferrous metal proportion $= \dfrac{\text{pounds of ferrous metals purchased}}{\text{pounds of total basic metals purchased}}$

of each category are listed in Table E-1. Within each of the four major industry groups, stainless steels are the most expensive ferrous metal, followed by steel castings, iron and steel forgings, alloy steels, iron castings, and then carbon steels. In general, the higher the unit cost of the ferrous metal, the more expensive it is to machine.[2]

The major categories of nonferrous metals used in SIC 34–37 and the average unit cost of each category are shown in Table E-2. In general, nonferrous metals which are comparably priced with ferrous metals are less expensive to machine.[3] For example, aluminum castings are comparably priced with steel castings but are much easier to machine, since aluminum generally has a much higher machinability rating than steel. Brass shapes (bars, plates, and pipes) are

2. There are several types of quantitative "machinability" indices which are used to compare the relative expense of cutting different types of metal. The most commonly used measures are based on (1) the allowable cutting speed for a 60 minute tool life, (2) the cubic volume of stock removed per minute for a 60 minute tool life, (3) the cutting tool life, in minutes, between sharpenings, (4) tool forces, energy, or power required, and (5) temperature of the cutting tool or chips. Typically, B112 hot-rolled (carbon) steel is taken as the standard of 100 percent machinability. If a stainless steel is rated at 50 percent, then only half as many inches of stainless steel could be removed with the same tool life. Correspondingly, if aluminum is rated at 500 percent, then five times as many inches of aluminum can be removed with the same tool life. An average machinability index could be computed for each category of ferrous metals in Table E-1 by averaging the individual machinability index for the many types of materials within each of the general categories listed. This average machinability index for each category could be compared with the unit cost to further substantiate the assertion that, in general, the difficulty of machining ferrous metals is related to the unit cost.

3. Several manufacturing engineers have pointed out that, since nonferrous metals are much more expensive than plain carbon steel, making mistakes much more costly, more care is typically taken in machining. If this were generally the case, the time required to process nonferrous inputs would be longer than would be indicated from a measure of a machinability index.

Table E-2: Average Unit Cost of Nonferrous Metal Inputs Used in SIC 34–37

Nonferrous Metal Input	Unit Cost (in dollars per pound)			
	34	35	36	37
Copper castings	1.30	1.47	1.49	1.81
Brass shapes	0.91	1.02	1.12	1.15
Aluminum forgings	1.04	—	—	0.84
Aluminum castings	0.63	1.10	0.97	1.03
Aluminum shapes	0.72	0.82	0.83	0.94
Refinery shapes	0.51	0.72	0.53	0.49
Scrap	0.43	—	—	0.08

Source: Derived from Tables E-3–E-6.

as highly priced as many types of stainless steel shapes, but are much less expensive to machine. Aluminum shapes are several times more expensive than carbon and alloy steel shapes, but are much less expensive to machine. If both ferrous and nonferrous metals are lumped together, there are problems with claiming that the unit cost of a basic metal indicates how difficult it is to machine.

Ideally, to represent the complexity of metal-shaping activities, there should be a separate unit cost measure for ferrous metals and for nonferrous metals. One could argue that the unit cost of each separate category of ferrous and nonferrous basic metal inputs, weighted by the quantity used, should be entered into the regression equation. Nonetheless, the average unit cost measure of basic metal inputs used here includes both ferrous and nonferrous metals.[4] However, as seen at the bottom of Table E-1, ferrous metals account for almost all (93 percent or more) of the quantity of basic metals purchased in SIC 34–37. For most industries in SIC 34–37, the inclusion of nonferrous metals in the average unit cost index of basic metals should result in only a small amount of distortion in the proposed index of the relative complexity of metal-shaping activities.

A more detailed breakdown of the categories of basic metal inputs purchased by the four major groups of metalworking industries is shown in Tables E-3 through E-6. The ratio of pounds of metal purchased to delivered cost is also shown in the tables. The last column in each table gives the *inverse* of unit cost, pounds per dollar. A small number of pounds per dollar indicates an expensive metal, whereas a large number of pounds per dollar indicates an inexpensive metal. For a given type of basic metal input, there are differences in the basic metal cost index across the major metalworking groups of industries. Carbon steel sheet and strip is the most widely used metal. On average in 1977, one dollar purchased 5.47 pounds of carbon steel sheet and strip in SIC 34, 5.27

4. The computations of the average unit cost of basic metal inputs were made before it was realized that ferrous metal inputs should be separated from nonferrous metal inputs.

Table E-3: Quantity and Cost of Basic Metal Inputs Used in SIC 34: Fabricated Metal Products

Basic Metal Input	Quantity Purchased (millions of pounds)	Delivered Cost (in millions of dollars)	Quantity/Cost (pounds/dollars)
Steel, total	87,106.2	17,818.8	4.89
Carbon steel			
Bars and shapes	8,196.0	1,518.9	5.40
Sheet and strip	37,229.4	6,806.2	5.47
Plates	6,701.4	1,149.8	5.83
Structural shapes	6,040.2	1,028.4	5.87
Wire and wire products	3,893.4	900.6	4.33
All other carbon steel			
mill shapes and forms, excluding rails	18,853.2	3,970.7	4.75
Alloy steel			
Bars and shapes	2,549.0	549.2	4.64
All other alloy steel mill shapes and forms	2,481.8	768.4	3.23
Stainless steel			
Sheet and strip	670.2	593.8	1.13
All other stainless steel shapes and forms	486.4	533.0	0.91
Bare wire	4.0	4.5	0.89
(for electrical conduction)			
Insulated wire	3.2	6.2	0.52
Brass mill shapes, total	839.8	760.1	1.10
Rod, bar, and mechanical wire	476.4	356.5	1.34
Plate, sheet, and strip	204.9	234.1	0.87
Pipe and tube	158.5	169.6	0.93
Aluminum and aluminum-based shapes, total	4,710.6	3,393.2	1.39
Sheet, plate, and foil	3,313.7	2,318.7	1.43
Extruded shapes	735.2	592.8	1.24
Other aluminum shapes	661.7	481.7	1.37
Castings, totala	2,097.8	834.2	2.51
Iron (gray and malleable)	1,373.0	322.0	4.26
Steel	465.0	272.4	1.71
Aluminum and aluminum-based	145.8	92.3	1.58
Copper and copper-based	114.0	147.5	0.77
Forgings, total	367.2	193.8	1.89
Iron and steel	358.4	184.6	1.94
Aluminum and aluminum-based	8.8	9.2	0.96
Metal Powder	—	—	—
Pig Iron	154.0	14.0	11.0
Nonferrous refinery shapes, total	937.8	477.0	1.97
Copper and copper-based	137.8	88.8	1.55
Zinc and zinc-based	306.0	114.3	2.68
Aluminum and aluminum-based	494.0	273.9	1.80
Scrap, total	2,064.0	116.8	17.67
Iron and steel scrap	1,979.8	80.4	24.62
Copper and copper-based alloy	38.0	20.2	1.88
Aluminum and aluminum-based alloy	46.2	16.2	2.85
Total basic metals	98,284.6	23,618.6	4.16

Source: Compiled from the 1977 Census of Manufactures (Bureau of the Census, 1981b,c,d).
—means data not provided in original source.
Note: Some categories may not add to totals shown because of rounding.
aRough and semifinished castings only (finished castings not included).

213

pounds in SIC 35, 5.33 pounds in SIC 36, and 5.97 pounds in SIC 37. One might suspect that industries which purchase the most pounds of metal pay lower prices per pound because of discounts for purchasing large volumes. This is not always the case. For example, for carbon steel sheet and strip in 1977, industries in SIC 34 purchased the most pounds, whereas industries in SIC 37 paid the lowest prices per pound. Other similar examples can be found in the tables. The fact that high-performance metals (those with special metallurgical properties) are more expensive suggests that, for a given basic metal input, differences in the cost index across industries are due at least in part to differences in the *quality* of the metal purchased.

For most types of ferrous and nonferrous metals, the industries in SIC 35, Machinery, except Electrical, pay more per pound than the industries in the other three major SIC groups considered. Industries in SIC 35 produce products which require high-performance metals, because they produce high-performance products such as precision machine tools, cutting tools, engines, turbines, and construction and farming equipment. It would be expected that for a given type of basic metal, industries in SIC 35 would require a higher quality, and hence more expensive, metal than the other industries. (The average number of pounds of basic metal per dollar is given for each major industry group at the bottom of Tables E-3 through E-6.) The aggregated index of pounds of basic metal per dollar is 4.16 for SIC 34 (Fabricated Metal Products), 3.60 for SIC 37 (Transportation Equipment), 3.26 for SIC 36 (Electrical Equipment and Machinery), and 2.76 for SIC 35 (Machinery, except Electrical). Based on a general understanding of the primary products manufactured within each major group, it appears that industries requiring metals with unusual structural properties purchase more expensive metals. This suggests there is a foundation for believing that the index of basic metal unit cost is an indicator of the complexity of the metal-shaping activities.

E.1.2. The Ratio of Processed Metal Cost to Basic Metal Cost

Inputs purchased by one metalworking industry from any industry within SIC 34–38, including itself, are referred to as processed metal inputs. The ratio of dollars of processed metals purchased to dollars of basic metals purchased varies greatly across the four major groups. The ratio ranges from a low value of 0.06 for SIC 34, where basic metals account for practically all of the metal input, to a high value of 4.04 for SIC 37, where basic metals account for less than a quarter of metal purchases. This ratio is also included as an explanatory factor in Equation E.1. I argue that the ratio can be interpreted as an indicator of the relative proportion of assembly to metal-shaping activities carried out within the industry. It is another way of measuring differences in the nature of the processing activities across industries. A processed metal input such as a screw, a stamping, or a motor will almost always require some type of assembly

Table E-4: Quantity and Cost of Basic Metal Inputs Used in SIC 35: Machinery, Except Electrical

Basic Metal Input	Quantity Purchased (in millions of pounds)	Delivered Cost (in millions) of dollars	Quantity/Cost (pounds/dollar)
Steel, total	26,467.6	6,688.3	3.96
Carbon steel			
Bars and shapes	4,856.4	1,163.7	4.17
Sheet and strip	6,896.0	1,309.4	5.27
Plates	6,204.4	1,163.7	5.33
Structural shapes	2,060.8	405.3	5.08
Wire and wire products	421.0	128.6	3.27
All other carbon steel mill shapes			
and forms, excluding rails	2,442.8	666.8	3.65
Alloy steel			
Bars and shapes	1,572.2	662.4	2.37
All other alloy steel mill shapes and forms	1,440.8	595.4	2.42
Stainless steel			
Sheet and strip	275.4	258.6	1.06
All other stainless steel shapes and forms	297.6	332.4	0.89
Bare wire	18.4	29.4	0.63
(for electrical conduction)			
Insulated wire	65.9	88.2	0.75
Brass mill shapes, total	493.0	501.1	0.98
Rod, bar, and mechanical wire	91.6	99.7	0.92
Plate, sheet, and strip	168.6	142.4	1.18
Pipe and tube	232.8	259.0	.90
Aluminum and aluminum-based shapes, total	709.8	582.5	1.22
Sheet, plate, and foil	372.1	305.1	1.22
Extruded shapes	162.2	141.3	1.15
Other aluminum shapes	175.5	136.1	1.29
Castings, total[a]	6,890.5	4,291.1	1.61
Iron (gray and malleable)	5,161.2	2,500.3	2.06
Steel	989.4	947.9	1.04
Aluminum and aluminum-based	650.6	712.6	0.91
Copper and copper-based	89.3	130.3	0.68
Forgings, total	1,895.0	1,211.1	1.56
Iron and steel	1,895.0	1,211.1	1.56
Aluminum and aluminum-based	—	—	—
Metal powder	149.2	262.1	0.57
Pig iron	161.6	16.0	10.1
Nonferrous refinery shapes, total	50.6	36.9	1.37
Copper and copper-based	50.6	36.9	1.37
Zinc and zinc-based	—	—	—
Aluminum and aluminum-based	—	—	—
Scrap, total	1,056.4	60.2	17.55
Iron and steel scrap	1,056.4	60.2	17.55
Copper and copper-based alloy	—	—	—
Aluminum and aluminum-based alloy	—	—	—
Total of basic metals	37,958.1	13,766.9	2.76

Source: Compiled from the 1977 Census of Manufactures (Bureau of the Census, 1981b).
—means data not provided in original source.
Note: Some categories may not add to totals shown because of rounding.
   [a]Rough and semifinished castings only (finished castings are not included).

215

Table E-5: Quantity and Cost of Basic Metal Inputs Used in SIC 36: Electrical and Electronic Equipment and Machinery

Basic Metal Input	Quantity Purchased (in millions of pounds)	Delivered Cost (in millions of dollars)	Quantity/Cost (pounds/dollar)
Steel, total	13,335.2	2,796.6	4.77
Carbon steel			
Bars and shapes	902.4	207.6	4.35
Sheet and strip	8,484.6	1,592.1	5.33
Plates	242.8	44.2	5.49
Structural shapes	278.2	34.5	8.06
Wire and wire products	724.0	147.0	4.92
All other carbon steel mill shapes and forms, excluding rails	1,611.4	311.9	5.17
Alloy steel			
Bars and shapes	24.0	11.1	2.16
All other alloy steel mill shapes and forms	891.6	285.3	3.12
Stainless steel			
Sheet and strip	119.8	93.6	1.28
All other stainless steel shapes and forms	56.4	69.2	0.81
Bare wire	102.3	118.1	0.87
(for electrical conduction)			
Insulated wire	227.3	341.5	0.67
Brass mill shapes, total	427.6	479.6	0.89
Rod, bar, and mechanical wire	220.2	229.7	0.96
Plate, sheet, and strip	162.8	196.5	0.83
Pipe and tube	44.6	53.4	0.84
Aluminum and aluminum-based shapes, total	571.1	471.2	1.21
Sheet, plate, and foil	246.2	225.1	1.09
Extruded shapes	146.9	133.7	1.10
Other aluminum shapes	178.0	112.4	1.58
Castings, totala	821.8	542.6	1.51
Iron (gray and malleable)	418.2	141.9	2.95
Steel	45.4	45.3	1.00
Aluminum and aluminum-based	326.7	316.4	1.03
Copper and copper-based	31.5	39.0	0.81
Forgings	16.0	9.2	1.74
Iron and steel	16.0	9.2	1.74
Aluminum and aluminum-based	—	—	—
Metal powder	164.3	30.8	5.33
Pig iron	—	—	—
Nonferrous refinery shapes, total	58.8	31.0	1.90
Copper and copper-based	23.2	17.9	1.30
Zinc and zinc-based	35.6	13.1	2.72
Aluminum and aluminum-based	—	—	—
Scrap, total			
Iron and steel scrap	—	—	—
Copper and copper-based alloy	—	—	—
Aluminum and aluminum-based alloy	—	—	—
Total of basic metals	15,724.4	4,820.6	3.26

Source: Compiled from the 1977 Census of Manufactures (Bureau of the Census, 1981b).
—means data not provided in original source.
Note: Some categories may not add to totals shown because of rounding.
  aRough and semifinished castings only (finished castings are not included).

Table E-6: Quantity and Cost of Basic Metal Inputs Used in SIC 37: Transportation Equipment

Basic Metal Input	Quantity Purchased (in millions of pounds)	Delivered Cost (in millions of dollars)	Quantity/Cost (pounds/dollar)
Steel, total	33,655.0	6,603.2	5.10
Carbon steel			
Bars and shapes	3,889.4	848.3	4.58
Sheet and strip	18,802.2	3,146.8	5.97
Plates	4,462.4	780.9	5.71
Structural shapes	1,546.4	297.0	5.21
Wire and wire products	424.8	131.7	3.22
All other carbon steel mill shapes and forms, excluding rails	1,622.8	405.3	4.00
Alloy steel			
Bars and shapes	1,704.0	441.9	3.86
All other alloy steel mill shapes and forms	782.2	242.6	3.22
Stainless steel			
Sheet and strip	316.6	212.9	1.49
All other stainless steel shapes and forms	104.0	95.9	1.08
Bare wire	27.0	35.4	0.76
(for electrical conduction)			
Insulated wire	99.3	113.5	0.87
Brass mill shapes, total	384.1	359.2	1.07
Rod, bar, and mechanical wire	183.0	144.9	1.26
Plate, sheet, and strip	165.9	174.4	0.95
Pipe and tube	35.2	39.9	0.88
Aluminum and aluminum based-shapes, total	993.9	936.5	1.06
Sheet, plate, and foil	638.7	585.8	1.09
Extruded shapes	312.1	304.6	1.02
Other aluminum shapes	43.2	46.1	0.94
Castings, total[a]	12,078.0	4,744.8	2.55
Iron (gray and malleable)	10,017.0	3,281.1	3.05
Steel	1,298.0	666.2	1.95
Aluminum and aluminum-based	750.6	774.8	0.97
Copper and copper-based	12.4	22.7	0.55
Forgings	3,564.6	1,641.8	2.17
Iron and steel	3,390.6	1,496.2	2.27
Aluminum and aluminum-based	174.0	145.6	1.19
Metal powder	211.6	68.8	3.08
Pig iron	97.0	8.9	10.90
Nonferrous refinery shapes	221.0	108.6	2.03
Copper and copper-based	—	—	—
Zinc and zinc-based	—	—	—
Aluminum and aluminum-based	221.0	108.6	2.03
Scrap, total	1,488.0	58.2	25.57
Iron and steel scrap	1,378.0	49.5	27.84
Copper and copper-based alloy	—	—	—
Aluminum and aluminum-based alloy	110.0	8.7	12.64
Total of basic metals	52,819.5	14,678.9	3.60

Source: Compiled from the 1977 Census of Manufactures (Bureau of the Census, 1981b).
—means data not provided in original source.
Note: Some categories may not add to totals shown because of rounding.
 [a]Rough and semifinished castings only (finished castings are not included).

217

or joining operation to integrate it with other parts of the product. In contrast, a basic metal input, such as carbon steel or aluminum or a semifinished forging will almost always require some type of metalcutting, metalforming, or finishing.

There are cases where this ratio might be an unreliable indicator of the relative proportion of assembly and joining to metal shaping and finishing within an industry. Purchased metal parts could require some additional metal shaping and finishing operations. There is a potentially more serious problem, in principle, in assuming that a low value for purchased metal parts indicates a small amount of assembly activity. If a product, such as a tractor, were to require a large mix of parts and the tractor producer were vertically integrated to a high degree, the producer might manufacture many of the needed parts and subassemblies from basic metal inputs. If this were the case, one must inquire how the high degree of vertical integration would be accomplished. If, say, one large establishment were to produce the tractor as its primary product and also produce all the accessory parts, the ratio of processed metal inputs to basic metal inputs would be low, since relatively few inputs would be purchased from other industries. This would accurately indicate that a large amount of metal shaping and finishing takes place within the industry, but it would inaccurately indicate that a minimal amount of assembly is required. If the tractor were assembled in one establishment, but major subcomponents, such as engines or screws, were produced in separate establishments, then these intrafirm transfers would still be counted as purchases from other industries in the *Census of Manufactures* data. In this case, the ratio of processed metal to basic metal inputs would accurately indicate the required relative amounts of assembly and joining versus metal shaping and finishing. Although some firms may be vertically integrated to a high degree, in my experience typical establishments are not. If it is the case that most subcomponents are produced in separate establishments, the ratio should generally be a reliable indicator of the relative proportions of assembly and joining versus machining and finishing. If the ratio of purchases of processed metal inputs to basic metal inputs is high, there is most certainly a large amount of assembly required within the industry. If the ratio is low, it is less certain whether this indicates that only a small amount of assembly activity is required.

Even with the acknowledged problems, the ratio of purchased metal products to basic metals seems to be a reasonable indicator of the amount of assembly activity. In SIC 34, where the ratio of processed metal to basic metal inputs is very small, assemblers make up only 11 percent of the production work force (see Table E-7). In SIC 35, where the ratio is close to unity, assemblers make up 16 percent of the production work force. In SIC 36 and 37, where the ratio is highest, assemblers make up 40 and 23 percent of the production workers, respectively. SIC 37 has the highest ratio, 4.04, but SIC 36 has the largest per-

Table E-7: Employment of Production Workers in SIC 34–37, Occupation and by Major Group, 1977

	34	35	36	37
Total production workers	1,191,810	1,368,450	1,201,160	1,330,380
Occupational Group	Percentage Distribution of Production Workers			
Fabrication (excluding assembly)	72	69	45	54
Assembly	11	16	40	23
Subtotal (fabrication plus assembly)	83	85	85	77
Inspection	3	4	5	6
Supervision	5	5	5	6
Maintenance and drivers	9	6	5	11
Cost of processed metal inputs to basic metal inputs	0.06	1.10	1.67	4.04

Source: Employment data compiled from Bureau of Labor Statistics, 1980.

centage of its production work force classified as assemblers. However, the ratio of 1.67 for SIC 36 probably underestimates the true value of the proportion of assembly to metal shaping activities, since most materials in SIC 36 are unspecified, and many of these unspecified materials are purchased parts.[5]

The ratio of processed metal costs to basic metal costs is substantially higher for the following industries than for all other industries: electronic computing equipment (SIC 3573), calculating and accounting machines (SIC 3574), radio and T.V. receiving equipment (SIC 3651), motor vehicles and equipment (3711), aircraft and parts (3721), and guided missiles and space craft (3761). All of these industries are known to require large amounts of assembly.

E.1.3. Hourly Wage Rates and Materials Coverage

Interindustry differences with respect to two other factors have previously been identified as potential sources of variation in unit cost across industries: the hourly wage of production workers and the proportion of total material costs that are covered by basic plus processed metal cost. A variable which directly corresponds to each of these factors can be constructed in a straightforward fashion. The "all included" average hourly wage of production workers is calculated using the ratio of wage and supplemental labor payments to hours

5. In the *1977 Census of Manufacturers* special report on *Selected Materials Consumed* (Bureau of the Census, 1981b), the results of a supplemental inquiry conducted in several selected industries are shown listing many of the materials which are not specified in the normal census survey on materials. Most of these materials are processed metal products. Since most of the industries included in the supplemental survey are in SIC 37, more complete data on the purchases of processed metal inputs are given for the industries in SIC 37 than for other industries. This partly accounts for the large ratio in SIC 37.

worked. The amount of total wage payments made to production workers for hours worked is given in the *1977 Census of Manufactures* (Bureau of the Census, 1981a,c,d). The amount of total supplemental payments to all wage and salaried workers is also given. The proportion of supplemental payments made to production workers is estimated by weighting total supplemental payments by the production labor share of total wage and salary payments. The median "all included" direct hourly wage is roughly $7 within each of the four major groups, ranging from $5 to $12. The ratio of basic plus processed metal cost to total material cost is used to compare the degree of material coverage across industries.

E.1.4. THE EFFECTS OF THE EXPLANATORY VARIABLES ON UNIT COST

The explanatory variables used in the multiple regression model are summarized in Table E-8. The proposed effect on unit cost is also shown by indicating whether the sign of the estimated elasticity of unit cost with respect to each variable should be positive or negative. The sign is determined by assuming that there is a 1 percent increase in the particular explanatory variable. The sign should be positive if a 1 percent increase in the explanatory variable should increase unit cost and negative if it should decrease unit cost. Unit cost should decrease as the level of output increases, so the sign of the coefficient for pounds of metal processed per establishment should be negative. The average unit cost of the basic metals purchased is called the basic metal cost index (bmci). The index gives the ratio of dollars per pound of basic metals purchased. An increase in the index means that more expensive metals are used, which is taken as an indication that the difficulty of the shaping operations increases. Since more difficult operations require more capital and/or labor inputs to accomplish, unit cost is assumed to increase and the proposed sign of this elasticity is positive.

The hourly wage of a bench assembler or an assembly machine operator is comparable to the hourly wage of many types of machine operators.[6] From this perspective, the degree of assembly should not affect unit cost. There are two reasons, however, for believing that a higher proportion of assembly to shaping indicates a more "complex" process. First, it is argued that the higher the ratio of processed metal inputs to basic metal inputs, the greater the diversity of material inputs used in an industry. Figure E-1 shows there is a tendency for the ratio of salary cost to production worker cost to increase as the ratio of processed metal inputs to basic metal inputs increases. This provides evidence for arguing

6. According to Ostwald (1981), in Chicago the hourly wage of a bench assembler in 1982 was $8.57, whereas the hourly wage for a machine operator ranged from $5.14 (drill press operator) to $11.60 (horizontal milling, drilling, boring-machine operator). (Ostwald notes: "The BLS data and other surveys are already out of date when published, and are inappropriate for estimating future costs. Therefore, these data are analyzed and projected to the mid-1982 period, and it is the time-adjusted values that are given [here]" [p. 3].)

Table E-8: Summary of Explanatory Variables in Multiple Regression Model

Factor	Variable	Notation	Effect on Unit Cost (Sign of Coefficient)
Level of output	$\dfrac{\text{pounds of metal processed}}{\text{number of establishments}}$	m/e	$\dfrac{d \ln(u)}{d\, l(m/e)} < 0$
Complexity of metal-shaping activities	$\dfrac{\text{dollars of basic metal}}{\text{pounds of basic metal}}$ = basic metal cost index	bmci	$\dfrac{d \ln(u)}{d \ln(bmci)} > 0$
Degree of assembly	$\dfrac{\text{dollars of purchased metal}}{\text{dollars of basic metal}}$	pmbm	$\dfrac{d\, l(u)}{d\, l(pmbm)} > 0$
"All included" hourly wage	$\dfrac{\text{production worker wages + benefits}}{\text{production worker hours}}$	w	$\dfrac{d\, l(u)}{d\, l(w)} > 0$
Material coverage	$\dfrac{\text{dollars of metals}}{\text{dollars of total materials}}$	c	$\dfrac{d\, l(u)}{d\, l(c)} < 0$

THE REGRESSION EQUATION IS
$$E[\ln(u)] = b_0 + b_1\ln(m/e) + b_2\ln(bmci) + b_3\ln(pmbm) + b_4\ln(w) + b_5\ln(c)$$
WHERE
$E[\ln(u)]$ = expected value of $l(u)$
u = unit cost
= value added/pounds of metal processed
b_i = ordinary least squares estimate of β_i

that it takes more organizational control and supervision to coordinate production when there is a larger proportion of processed metal inputs. Second, the inspection of assembled products requires more than just the verification of dimensions. Since subcomponents must be properly integrated with one another, testing is required to verify that the final product performs its designated functions properly. With electronics equipment and computers this can be a fairly extensive and complicated process. A breakdown of the percentage of employment by major occupational groups is shown in Table E-7. The two major industry groups with the largest ratio of processed metal inputs to basic metal inputs are SIC 36 and SIC 37. These two industries have the largest proportion of assemblers, as previously noted, but they also have the largest proportion of inspectors. Thus, there is some evidence for arguing that a greater proportion of processed metal products corresponds to products which are more complex in the sense that more subcomponents have to be integrated and tested. Given these assumptions, unit cost should increase with an increase in the ratio of dollars of processed metals to dollars of basic metals, and the elasticity of unit cost with respect to output should be positive.

An increase in the "all included" wage rate increases the production labor cost of performing an hour's work, which increases value added. The wage rate

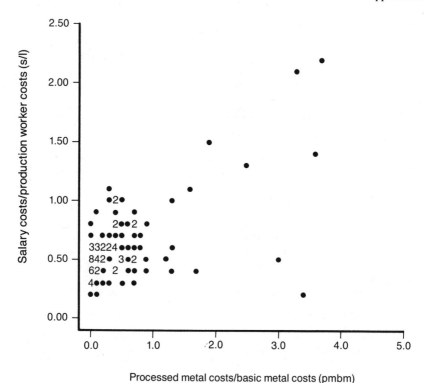

THE REGRESSION EQUATION IS
(s/l) = 0.488 + 0.238 ln(pmbm)

Variable	Coefficient	St. dev. of coef.	T-ratio = coef./S.D.
constant	0.48794	0.03268	14.93
ln(pmbm)	0.23777	0.03366	7.06

The st. dev. of Y about regression line is S = 0.2673 with (101 – 2) = 99 degrees of freedom
R^2 = 33.5 percent
R^2 = 32.8 percent, adjusted for d.f.

Figure E-1: Ratio of Salaried Workers to Production Workers versus Ratio of Cost of Processed Metal Inputs to Cost of Basic Metal Inputs

should have no effect on the level of output, so the sign of the elasticity of unit cost with respect to wages should be positive. Figure E-2 shows that the "all included" hourly wage tends to increase as the level of output per establishment increases. This strong positive correlation between the hourly wage rate and the level of output will introduce some degree of multicollinearity into the regression and may complicate the processes of determining the separate influences of the output variable and the wage variable.

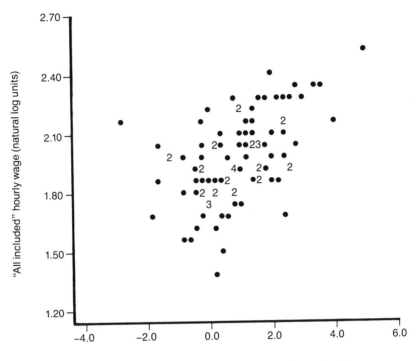

Pounds of metal processed/establishment (m/e)
(natural log units)

THE REGRESSION EQUATION IS
ln(w) = 1.90 + 0.0815 ln(m/e)

Variable	Coefficient	St. dev. of coef.	T-ratio = coef./S.D.
constant	1.89693	0.02231	85.03
ln(m/e)	0.08147	0.01406	5.80

The st. dev. of Y about regression line is S = 0.1832 with (101 – 2) = 99 degrees of freedom
R^2 = 25.3 percent
R^2 = 24.6 percent, adjusted for d.f.

Figure E-2: Hourly Wage of Production Workers versus Output

Material coverage is given by the ratio of basic and processed metal costs to total material costs. As the ratio increases, capital and labor costs, which compose value added, become more proportional to the processing of the basic and processed metals which are included in the measure of output. If the ratio is low, then the processing of materials which are not included in the output measure contributes to the value added. The elasticity of unit cost with respect to coverage should be negative, since both value added and processing of materials which are unaccounted for in the output measure increase as coverage decreases.

E.2. Regression Results

The results of regressing value added per pound of metal against the five explanatory variables for 101 of the four-digit industries in SIC 34–37 are summarized in Table E-9. The comparable regression results for each of the four major groups of industries, SIC 34, SIC 35, SIC 36, and SIC 37, are summarized in Tables E-10, E-11, and E-12 and E-13, respectively. The discussion of the results proceeds by addressing the following questions for each set of regressions:

— Is there a relation between the dependent variable and the full set of explanatory variables?

— How much of the variance in unit cost is explained by variables in the model?

— Are the estimates of the coefficients "significant" enough to allow for a meaningful comparison of the relative influence of each factor?

— How do the signs of the estimated coefficients compare with the predictions of the signs in Table E-8?

— What can be said about the relative influence of the level of output versus the nature of the processing activities on unit cost?

The first two items are straightforward and do not require much elaboration, assuming the reader is familiar with the basics of regression models. To test whether there is a relation between the dependent variable, $\ln(u)$, and the set of five explanatory variables, $\ln(m/e)$, $\ln(bmci)$, $\ln(pmbm)$, $\ln(w)$, and $\ln(c)$, is to choose between the two alternatives:

$$H_0: \beta_1 = \beta_2 = \beta_3 = \beta_4 = \beta_5 = 0$$
$$H_1: \text{not all } \beta_k \ (k = 1, \ldots, 5) = 0$$

The test statistic is given by

$$F^* = \frac{\text{regression mean square}}{\text{error mean square}} = \frac{\text{MSR}}{\text{MSE}}$$

The decision rule is as follows when the risk of Type I error (the probability of incorrectly rejecting the maintained hypothesis when it is true) is to be controlled at level a:

$$\text{If } F^* < F(1 - a; p - 1, n - p), \text{ conclude } H_0$$
$$\text{If } F^* > F(1 - a; p - 1, n - p), \text{ conclude } H_1$$

where

$F(1 - a; p - 1, n - 1) = $ the value of the F statistic with $p - 1$ and $n - 1$ degrees of freedom, which cuts of $1 - a$ of the area of the distribution in the tail

Table E-9: Regression Results for Total Four-Digit Data Set: SIC 34–37

Output: pounds of metal /number of establishments
THE REGRESSION EQUATION IS
$\ln(u) = -1.17 - 0.295 \ln(m/e) + 0.983 \ln(bmci) + 0.488 \ln(1 + pmbm)$
$\qquad + 0.948 \ln(w) - 0.765 \ln(c)$

VARIABLE	COEFFICIENT	ST. DEV. OF COEF.	T-RATIO = COEF./S.D.
constant	−1.1707	0.4625	−2.53
ln(m/e)	−0.2947	0.0387	−7.61
ln(bmci)	0.9827	0.0909	10.81
ln(1 + pmbm)	0.4881	0.0490	9.96
ln(w)	0.09478	0.2151	4.41
ln(c)	−0.7644	0.2051	−3.73

THE ST. DEV. OF Y ABOUT REGRESSION LINE IS
$S = 0.3717$
WITH $(101 - 7) = 95$ DEGREES OF FREEDOM
$R^2 = 87.0$ PERCENT
$R^2 = 86.3$ PERCENT, ADJUSTED FOR D.F.
ANALYSIS OF VARIANCE

DUE TO	DF	SS	MS = SS/DF
REGRESSION	5	87.8546	17.5709
RESIDUAL	95	13.1252	0.1382
TOTAL	100	100.9799	

DEGREE OF MULTICOLLINEARITY: Regress ln(m/e) against other explanatory variables.
THE REGRESSION EQUATION IS
$\ln(m/e) = -5.68 - 1.14 \ln(bmci) + .0630 \ln(1 + pmbm) + 2.83 \ln(w) + 0.448 \ln(c)$

VARIABLE	COEFFICIENT	ST. DEV. OF COEF.	T-RATIO = COEF./S.D.
constant	−5.679	1.072	−5.30
ln(bmci)	−1.1382	0.2096	−5.43
ln(1 + pmbm)	0.0630	0.1290	0.49
ln(w)	2.8275	0.4879	5.79
ln(c)	0.4479	0.5385	0.83

THE ST. DEV. OF Y ABOUT REGRESSION LINE IS
$S = 0.9794$
WITH $(101 - 6) = 96$ DEGREES OF FREEDOM
$R^2 = 45.8$ PERCENT
$R^2 = 43.5$ PERCENT, ADJUSTED FOR D.F.
CORRELATIONS AMONG EXPLANATORY VARIABLES

	ln(m/e)	ln(bmci)	ln(1 + pmbm)	ln(w)
ln(bmci)	−0.471			
ln(1 + pmbm)	−0.015	0.254		
ln(w)	0.503	0.050	0.129	
ln(c)	0.313	−0.270	−0.010	0.277

p = the number parameters estimated = 6
n = the number of observations

In all cases (in the combined four data set of all four major groups, and in the separate data set for each of the four major groups), the value of the test statistic, F^*, is large, indicating that a strong relation between the dependent variable and the set of explanatory variables does exist. The coefficient of multiple determination, R^2, measures the proportionate reduction of the total variation in the dependent variable associated with the use of the set of explanatory variables. R^2 measures the proportion of variation in ln(u) that is explained or accounted for by differences in the levels of the five variables in the model. It is defined as follows:

$$R^2 = 1 - \frac{\text{error sum of squares}}{\text{total sum off squares}} = 1 - \frac{SSE}{SSTO}$$

When adjusted for the number of explanatory variables included in the model, the definition is revised to

$$R^2, \text{adjusted} = 1 - (\frac{n-1}{n-p})\frac{SSE}{SSTO}$$

where

$n - 1$ = degrees of freedom for the total sum of squares

$n - p$ = degrees of freedom for the error sum of squares

The five explanatory variables explain just under 90 percent of the variation in ln(u) (using R^2, adjusted) within the combined data set. They also explain close to 90 percent of the variation in SIC 34, 35, and 36, and slightly less than 80 percent of the variation in SIC 37. The fact that the levels of these variables explain most of the variation in unit cost deserves a moment's worth of reflection. There are a great number of specific techniques used and distinct product types manufactured within the metalworking sector. Yet, despite these basic differences, there still appears to be some type of underlying structure which explains why unit cost varies across these industries. This structure is apparent across the entire pooled sample of observations, and also within each of the four major groups of industries (SIC 34–37).

In a statistical sense, a coefficient is significant if its standard deviation is small in comparison with its estimated magnitude. Whether or not an estimated coefficient is significant is usually determined from the result which states that

$$\frac{b_k - \beta_k}{s(b_k)} \text{ is distributed as } t(n - p)$$

assuming the assumptions of the normal error model are not violated (that is, the error terms for each observation are independent and normally distributed with zero mean and a constant variance)

where

b_k = the estimated value of the coefficient β_k

β_k = the true (but unknown) value of the coefficient

$s(b_k)$ = the standard deviation of the estimate of b_k

Because the quantity above has a t distribution, an interval can be defined such that the probability that the "true" value of the parameter lies within the interval is $1 - a$. A $1 - a$ confidence interval for the parameter β_k is given by

$$b_k - t(1 - \frac{a}{2};n - p)s(b_k) < \beta_k < b_k + t(1 - \frac{a}{2};n - p)s(b_k) \qquad \text{(E.3)}$$

where

$t(1 - \frac{a}{2}; n - p)$ = value of the t statistic with $n - p$ degrees of freedom, which cuts off $a/2$ of the area of the distribution in each of the tails

The smaller the standard deviation of the estimate, the narrower the interval and the more certainty there is about the magnitude of the parameter. If the standard deviation is large relative to the magnitude of the estimate, then the interval is very wide, and the estimate is considered to be imprecise. If the interval includes zero, then it is not possible to know if the coefficient is either positive or negative. When the confidence interval includes zero, the coefficient is considered to be insignificant, since it cannot be said with $1 - a$ probability that the estimate is significantly different from zero.

Perfect multicollinearity is the case where one explanatory variable is perfectly correlated with another explanatory variable *or* with a linear combination of other explanatory variables. When there is an exact linear relation among the explanatory variables, it is not possible to solve the multiple linear regression model and obtain estimates for the coefficients. If there is a high degree of multicollinearity there is a strong, but not perfect, correlation between one explanatory variable and another or with a linear combination of other explanatory variables. With a high degree of multicollinearity, solutions for the coefficients can still be obtained, but the estimates for the correlated explanatory variables have large variances, and are therefore imprecise. If the degree of multicollinearity were strong enough, it would not be possible to estimate the separate influences of each of the explanatory variables, even when the set of variables accounts for almost all of the variation in the dependent variable. In an extreme case, each of the individual coefficients could be statistically insignificant, even though there is a definite regression relationship between the dependent variable and the set of explanatory variables. One means of measuring the degree of linear dependence among explanatory variables is to regress one against the others. In the lower half of the Tables E-9 through E-13 summarizing the regression results, pounds of metal is regressed against the four other explanatory variables. In two of the major groups, SIC 34 and SIC 36, nearly 70 percent of the variation in output per establishment is explained by a linear combination of the four other explanatory variables. This correlation is strong enough to raise some doubts about the precision of the estimates in these two cases. Within the pooled data set, and within the two other major groups, SIC 35 and SIC 37, the regression of pounds of metal per establishment against the

Table E-10: Regression Results for SIC 34: Fabricated Metal Products

THE REGRESSION EQUATION IS

$\ln(u) = -1.34 - 0.298 \ln(m/e) + 0.893 \ln(bmci) + 0.789 \ln(1 + pmbm)$
$\quad + 0.845 \ln(w) - 1.29 \ln(c)$

VARIABLE	COEFFICIENT	ST. DEV. OF COEF.	T-RATIO = COEF./S.D.
constant	−1.3420	0.8057	−1.67
ln(m/e)	−0.2977	0.0891	−3.34
ln(bmci)	0.8933	0.1408	6.34
ln(1 + pmbm)	0.7893	0.4346	1.82
ln(w)	0.8445	0.4241	1.99
ln(c)	−1.2887	0.3989	−3.23

THE ST. DEV. OF Y ABOUT REGRESSION LINE IS
$S = 0.2429$
WITH (28 − 6) = 22 DEGREES OF FREEDOM
$R^2 = 91.5$ PERCENT
$R^2 = 89.6$ PERCENT, ADJUSTED FOR D.F.
ANALYSIS OF VARIANCE

DUE TO	DF	SS	MS = SS/DF
REGRESSION	5	13.96849	2.79370
RESIDUAL	22	1.29845	0.05902
TOTAL	27	15.26694	

DEGREE OF MULTICOLLINEARITY: Regress ln(m/e) against other explanatory variables.
THE REGRESSION EQUATION IS

$\ln(m/e) = -6.72 - 0.722 \ln(bmci) - 0.659 \ln(1 + pmbm) + 3.83 \ln(w) + 1.224 \ln(c)$

VARIABLE	COEFFICIENT	ST. DEV. OF COEF.	T-RATIO = COEF./S.D.
constant	−6.176	1.263	−5.32
ln(bmci)	−0.7216	0.2933	−2.46
ln(1 + pmbm)	−0.659	1.008	−0.65
ln(w)	3.8316	0.5982	6.50
ln(c)	1.2249	0.8981	1.36

THE ST. DEV. OF Y ABOUT REGRESSION LINE IS
$S = 0.5687$
WITH (28 − 5) = 23 DEGREES OF FREEDOM
$R^2 = 73.6$ PERCENT
$R^2 = 69.0$ PERCENT, ADJUSTED FOR D.F.
CORRELATION AMONG EXPLANATORY VARIABLES

	ln(m/e)	ln(bmci)	ln(1 + pmbm)	ln(w)
ln(bmci)	−0.473			
ln(1 + pmbm)	−0.084	−0.044		
ln(w)	0.723	−0.059	−0.039	
ln(c)	0.381	−0.590	−0.003	0.008

Table E-11: Regression Results for SIC 35: Machinery, except Electrical

THE REGRESSION EQUATION IS

$\ln(m/e) = -10.2 - 1.24 \ln(bmci) + 0.358 \ln(pmbm) + 4.78 \ln(w) - 0.262 \ln(c)$
$\qquad + 0.838 \ln(w) - 0.094 \ln(c)$

VARIABLE	COEFFICIENT	ST. DEV. OF COEF.	T-RATIO = COEF./S.D.
constant	0.0043	0.7182	0.01
ln(m/e)	-0.2725	0.0450	-6.05
ln(bmci)	1.2374	0.1116	11.09
ln(pmbm)	0.1905	0.0308	6.18
ln(w)	0.8378	0.3438	2.44
ln(c)	-0.0944	0.2378	-0.40

THE ST. DEV. OF Y ABOUT REGRESSION LINE IS

S = 0.2255

WITH (41 – 6) = 35 DEGREES OF FREEDOM

R^2 = 90.2 PERCENT

R^2 = 88.8 PERCENT, ADJUSTED FOR D.F.

ANALYSIS OF VARIANCE

DUE TO	DF	SS	MS = SS/DF
REGRESSION	5	16.38817	3.27763
RESIDUAL	35	1.77959	0.05085
TOTAL	40	18.16776	

DEGREE OF MULTICOLLINEARITY: Regress ln(m/e) against other explanatory variables.

THE REGRESSION EQUATION IS

$\ln(m/e) = -10.2 - 1.24 \ln(bmci) + 0.358 \ln(pmbm) + 4.78 \ln(w) - 0.262 \ln(c)$

VARIABLE	COEFFICIENT	ST. DEV. OF COEF.	T-RATIO = COEF./S.D.
constant	-10.216	2.041	-5.01
ln(bmci)	-1.2425	0.3571	-3.48
ln(pmbm)	0.3583	0.09710	3.69
ln(w)	4.7844	0.9913	4.83
ln(c)	-0.2623	0.8789	-0.30

THE ST. DEV. OF Y ABOUT REGRESSION LINE IS

S = 0.8344

WITH (41 – 5) = 36 DEGREES OF FREEDOM

R^2 = 53.7 PERCENT

R^2 = 48.6 PERCENT, ADJUSTED FOR D.F.

CORRELATION AMONG EXPLANATORY VARIABLES

	ln(m/e)	ln(bmci)	ln(pmbm)	ln(w)
ln(bmci)	-0.401			
ln(pmbm)	0.289	-0.046		
ln(w)	0.425	0.024	-0.319	
ln(c)	-0.070	-0.095	-0.167	0.003

Table E-12: Regression Results for SIC 36: Electrical and Electronic Machinery and Equipment

THE REGRESSION EQUATION IS

$\ln(u) = 0.132 - 0.230 \ln(m/e) + 1.20 \ln(bmci) + 0.241 \ln(pmbm)$
$\quad + 0.914 \ln(w) - 0.0870 \ln(c)$

VARIABLE	COEFFICIENT	ST. DEV. OF COEF.	T-RATIO = COEF./S.D.
constant	0.1317	0.9161	0.14
ln(m/e)	−0.2300	0.0842	−2.73
ln(bmci)	1.1955	0.1702	7.03
ln(pmbm)	0.2405	0.0545	4.41
ln(w)	0.9140	0.4167	2.19
ln(c)	−0.0870	0.5408	−0.16

THE ST. DEV. OF Y ABOUT REGRESSION LINE IS

S = 0.2637

WITH (18 − 6) = 12 DEGREES OF FREEDOM

R^2 = 93.5 PERCENT

R^2 = 93.6 PERCENT, ADJUSTED FOR D.F.

ANALYSIS OF VARIANCE

DUE TO	DF	SS	MS = SS/DF
REGRESSION	5	17.54410	3.50882
RESIDUAL	12	0.83462	0.06955
TOTAL	17	18.37872	

DEGREE OF MULTICOLLINEARITY: Regress ln(m/e) against other explanatory variables.

THE REGRESSION EQUATION IS

$\ln(m/e) = -6.30 - 1.27 \ln(bmci) - 0.281 \ln(pmbm) + 3.12 \ln(w) - 0.347 \ln(c)$

VARIABLE	COEFFICIENT	ST. DEV. OF COEF.	T-RATIO = COEF./S.D.
constant	−6.297	2.462	−2.56
ln(bmci)	−1.2684	0.4366	−2.91
ln(pmbm)	−0.2809	0.1619	−1.73
ln(w)	3.119	1.066	2.93
ln(c)	−0.347	1.779	−0.20

THE ST. DEV. OF Y ABOUT REGRESSION LINE IS

S = 0.8690

WITH (18 − 5) = 13 DEGREES OF FREEDOM

R^2 = 70.3 PERCENT

R^2 = 61.1 PERCENT, ADJUSTED FOR D.F.

CORRELATION AMONG EXPLANATORY VARIABLES

	ln(m/e)	ln(bmci)	ln(pmbm)	ln(w)
ln(bmci)	−0.675			
ln(pmbm)	−0.289	0.120		
ln(w)	0.589	−0.276	0.088	
ln(c)	0.483	0.549	−0.278	0.319

Table E-13: Regression Results for SIC 37: Transportation Equipment

THE REGRESSION EQUATION IS

$Y = -4.05 - 0.604 \ln(m/e) + 0.298 \ln(bmci) + 0.365 \ln(pmbm) + 2.53 \ln(w) + 0.758 \ln(c)$

VARIABLE	COEFFICIENT	ST. DEV. OF COEF.	T-RATIO = COEF./S.D.
constant	−4.052	1.726	−2.35
ln(m/e)	−0.6039	0.1523	−3.97
ln(bmci)	0.2983	0.3116	0.96
ln(pmbm)	0.3646	0.1337	2.73
ln(w)	2.5320	0.8639	2.93
ln(c)	0.7576	0.7582	0.96

THE ST. DEV. OF Y ABOUT REGRESSION LINE IS

$S = 0.6106$

WITH (14 − 6) = 8 DEGREES OF FREEDOM

$R^2 = 86.2$ PERCENT

$R^2 = 77.5$ PERCENT, ADJUSTED FOR D.F.

ANALYSIS OF VARIANCE

DUE TO	DF	SS	MS = SS/DF
REGRESSION	5	18.5881	3.7176
RESIDUAL	8	2.9829	0.3729
TOTAL	13	21.5709	

DEGREE OF MULTICOLLINEARITY: Regress ln(m/e) against other explanatory variables.

THE REGRESSION EQUATION IS

$\ln(m/e) = -6.28 - 0.900 \ln(bmci) - 0.102 \ln(pmbm) + 3.54 \ln(w) + 1.06 \ln(c)$

VARIABLE	COEFFICIENT	ST. DEV. OF COEF.	T-RATIO = COEF./S.D.
constant	−6.277	3.146	−1.99
ln(bmci)	−0.9004	0.6126	−1.47
ln(pmbm)	−0.1022	0.2906	−0.35
ln(w)	3.545	1.477	2.40
ln(c)	1.064	1.682	0.63

THE ST. DEV. OF Y ABOUT REGRESSION LINE IS

$S = 1.357$

WITH (14 − 5) = 9 DEGREES OF FREEDOM

$R^2 = 54.6$ PERCENT

$R^2 = 34.5$ PERCENT, ADJUSTED FOR D.F.

CORRELATIONS AMONG EXPLANATORY VARIABLES

	ln(m/e)	ln(bmci)	ln(pmbm)	ln(w)
ln(bmci)	−0.263			
ln(pmbm)	0.178	0.232		
ln(w)	0.646	0.091	0.457	
ln(c)	0.301	0.329	0.332	0.434

other explanatory variables shows that the degree of multicollinearity is not severe, and should not, therefore, cause serious problems with the precision of the estimates.

Earlier in Chapter 4, the question was raised whether the level of output or the nature of the processing activities exerted a greater influence on unit cost. Within the framework developed so far, the way to answer the question is to examine the magnitudes of the parameter estimates for pounds of metal per establishment (β_1), the basic metal cost index (β_2), and the proportion of assembly to metal-shaping activities (β_3). Given that all parameters are elasticities, the larger the absolute magnitude of the estimate, the greater the percentage of change in unit cost with a 1 percent change in the respective explanatory variable. One must proceed cautiously when comparing magnitudes of these estimates in order to determine which factors exert a greater influence on unit cost. The definition of the confidence interval previously given in Equation E.3 must be adjusted when estimates of several coefficients in a multiple regression are of interest, in order to ensure that there is a $1 - a$ probability that *all* of the parameters of interest lie within their designated intervals. For each set of regressions, Bonferroni joint confidence intervals are specified for the parameters for the pounds of metal, the basic metal cost index, and the ratio of processed metal cost to basic metal cost, and they are defined as follows:

$$b_1 - Bs(b_1) < \beta_1 < b_1 + Bs(b_1)$$
$$b_2 - Bs(b_2) < \beta_2 < b_2 + Bs(b_2)$$
$$b_3 - Bs(b_3) < \beta_3 < b_3 + Bs(b_3)$$

where

$$B = t(1 - \frac{a}{2s}; n - p)$$

s = number of confidence statements = 3
n = number of observations
p = number of parameters estimated in regression = 6
β_1 = parameter for pounds of metal processed (m)
β_2 = parameter for basic metal cost index (bmci)
β_3 = parameter for ratio of processed metal cost to basic metal cost(pmbm)
b_i = estimate of parameter β_i, i = 1,2,3 s(b_i)
    = standard deviation of b_i, i = 1,2,3

Reiterating, using the Bonferroni method, there is *at least* a $1 - a$ probability that β_1 lies within its interval and β_2 lies within its interval and β_3 lies within its interval. The joint confidence intervals for these three parameters are shown for the five sets of regressions in Tables E-14 and E-15. Only these three parameters are examined in detail, because the issue under investigation here is the relative influence of the level of output and the nature of the processing requirements on unit cost.

Table E-14: Bonferroni Joint Confidence Intervals for Output and Complexity Parameters, Total Data Set SIC 34–37, n = 101

	Parameter for:	
ln(m/e)	ln(bmci)	ln(1 + pmbm)
$-0.3867 < \beta_1 < -0.2029$	$0.7669 < \beta_2 < 1.1985$	$0.3718 < \beta_3 < 0.6044$
$b_1 = -0.2948$	$b_2 = 0.9847$	$b_3 = 0.4881$

Family Confidence Coefficient = (1.0 − 0.05) − 0.95

See Table 4-17 for definitions of variables and full regression model.

E.2.1. REGRESSION RESULTS FOR THE TOTAL DATA SET, SIC 34–37

The results obtained for the 101 observations in the total data set are shown in Table E-9. The regression model explains nearly 87 percent of the variation in unit cost. There is some degree of multicollinearity among the explanatory variables. The number of pounds of metal processed per establishment is correlated with the basic metal cost index and the hourly wage. However, the regression of pounds of metal processed per establishment on the rest of the explanatory variables in the bottom of the table indicate that the degree of multicollinearity is not high. Also, the large t-ratios of the estimates, especially for the output and complexity parameters, indicates that the separate influences of each parameter can still be measured meaningfully. The confidence intervals for the output and complexity parameters are shown in Table E-14. The probability that all three

Table E-15: Bonferroni Joint Confidence Intervals for Output and Complexity Parameters, Separated for Major Industry Groups SIC 34–37

Data Set	Parameter for		
	l(m/e)	l(bmci)	l(pmbm)[a]
SIC 34 n = 28	$-0.5212 < \beta_1 < -0.0742$	$0.5402 < \beta_2 < 1.2464$	$-0.3007 < \beta_3 < 1.8793$
	$b_1 = -0.2977$	$b_2 = 0.8933$	$b_3 = 0.7893$
SIC 35 n = 41	$-0.3845 < \beta_1 < -0.1645$	$0.9651 < \beta_2 < 1.5097$	$0.1153 < \beta_3 < 0.2657$
	$b_1 = -0.2725$	$b_2 = 1.2374$	$b_3 = 0.1905$
SIC 36 n = 18	$-0.4557 < \beta_1 < -0.0043$	$0.7392 < \beta_2 < 1.6518$	$0.0944 < \beta_3 < 0.3866$
	$b_1 = -0.2300$	$b_2 = -1.1955$	$b_3 = 0.2405$
SIC 37 n = 14	$-1.0450 < \beta_1 < -0.1628$	$-0.6041 < \beta_2 < 1.2007$	$-0.0226 < \beta_3 < 0.7518$
	$b_1 = -0.6039$	$b_2 = 0.2983$	$b_3 = 0.3646$

Family Confidence Coefficient for Each Major Industry Group = (1 − 0.05) = 0.95

See Table E-8 for definitions of variables and for full regression model.
Intervals are derived from regression results in Tables E-10 through E-13.
[a] In SIC 34, the variable is ln(1 + pmbm).

parameters lie within the designated intervals is 95 percent. All three parameters are highly significant.

The estimate of the coefficient for the basic metal cost index, b_2, is larger in absolute value than the estimate for b_1. The absolute values of the joint confidence intervals for β_1 and β_2 do not overlap, so it is possible to conclude, with a high probability, that a 1 percent change in the basic metal cost index exerts more influence on unit cost than a 1 percent change in the number of pounds of metal processed. Given the interpretations attached to these variables, this is rephrased in the following way: The complexity of the metal-shaping activities appears to exert a stronger influence on unit cost than does the level of output produced.

Why is this so? The basic metal cost index is plotted against pounds of metal processed in Figure E-3. Across the metalworking industries, the unit cost of the basic metals purchased tends to decrease as the number of pounds of metal processed per establishment increases. This indicates that low-volume producers tend to use more expensive basic metal inputs, and that high-volume producers tend to use lower cost, more standardized material inputs. The tendency to change the types of material input as the level of output increases suggests that increasing the scale of production means more than just increasing the volume of production. In their analysis of the bicycle headlug, Lamyai, Rhee, and Westphal (1978: 32) point out that "more highly mechanized systems use raw materials having lower unit prices, which suggests that they substitute for further processing in the raw material supplying the industry." It is noteworthy that this same effect is captured in the cross-sectional data set of metalworking industries. One plausible explanation for the significance of the decrease in unit basic metal cost with increasing output is that lower cost (and more standard) material inputs are suggestive of simplified and standardized product designs as well as of simplified and standardized processing requirements. Giving more attention to "design for manufacturability" when products are made in large volumes is a well-established practice in manufacturing. The cost of the material and the ease with which it can be processed are obviously of greater concern in repetitive processing than when only a few copies of a product are produced.

There is no a priori reason for believing that batch-produced products are intrinsically more complex than mass-produced products. For example, while an automobile is probably not as complex as a jet fighter aircraft, it is probably comparable in complexity to many specially designed machines produced in small volumes. The total number of units produced (or the average batch size) accounts for only part of the reason low-volume production is more expensive than high-volume production. The complexity of the operations which must be carried out on the basic material inputs (which are directly related to the design of the product) appears to explain a large part of the difference in unit cost between low- and high-volume production.

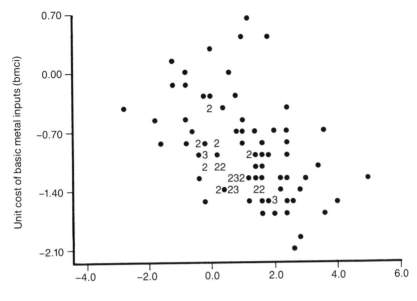

Pounds of metal processed per establishment (m/e)

(Natural log units)

THE REGRESSION EQUATION IS

ln(bmci) = −0.852 − 0.1821 ln(m/e)

Variable	Coefficient	St. dev. of coef.	T-ratio = coef./S.D.
constant	0.85247	0.05431	−15.70
ln(m/e)	−0.18165	0.03422	−5.31

The st. dev. of Y about regression line is S = 0.4460 with (101 − 2) = 99 degrees of freedom

R^2 = 22.2 percent

R^2 = 21.4 percent, adjusted for d.f.

Figure E-3: Unit Cost of Basic Metal Inputs versus Output: Pooled Four-Digit Data Set

Suppose changes in the types of material used (or more broadly, the optimiza-tion of product design and processing requirements for ease of manufac-turability) do account for a substantial part of the differential in unit cost between low-volume and high-volume manufacturing. If the level of output in a batch-production factory were increased to the level of a mass production fac-tory *without* altering the types of material used or the nature of the processing requirements, then unit cost in the high-volume batch-production facility should still be higher than in the mass-production facility. This might provide an explanation for why the regression equation which includes a variable repre-senting the nature of the processing requirements, in addition to an output vari-able, gives a smaller estimate of the magnitude of the output elasticity than regression equations without such a variable (See Table 4-19). Without an explanatory variable to incorporate the changes in the types of material used (and in the way the product is made), the output elasticity incorporates the effects of increasing batch size *and* of altering the nature of the processing requirements. The presence of such an explanatory variable, would incorporate some of the effects of altering the material inputs used, which would reduce the influence of the output elasticity. Output elasticity estimated from the cross-sectional data, without including the other explanatory variables (Case B in Table 4-19), is bounded by the Lamyai, Rhee, and Westphal (1978) estimates, which are designed to encompass the lowest-cost and the highest-cost estimates for producing the product they analyzed. However, when the basic metal cost index and the other explanatory variables are included in the cross-sectional analysis, the influence of the scale parameter is lessened and its magnitude is less than that of the lower Lamyai, Rhee, and Westphal estimate.

The confidence interval for β_3 indicates that unit cost also increases as the ratio of processed metal cost to basic metal cost increases, with all other factors held constant. This suggests that unit cost increases as the relative proportion of assembly to metal shaping increases. With the basic metal cost index held constant, the ratio of assembly to metal-shaping activities appears to have an influence on unit cost comparable with that of output. The ratio of the processed metal input cost to basic metal input cost is plotted against pounds of metal processed in Figure E-4. Although unit cost has been shown to increase as the ratio of processed metal inputs to basic metal inputs increases, there is no discernible pattern between the measure of the degree of assembly and the level of output. With one exception, the industries with the highest levels of assembly are not large-volume producers.

E.2.2. REGRESSION RESULTS FOR SIC 34

The regression results for each of the separate major industry groups, SIC 34, 35, 36, and 37, are discussed in turn. The reader is referred Tables E-10 through E-13, which summarize these regression models, as well as to Table E-15, which gives the joint confidence intervals for the output parameter and for the two complexity parameters. The explanatory variables in SIC 34 account for nearly 90 percent of the variation in unit cost. There is a high degree of multicollinearity among the explanatory variables, due largely to the high positive correlation between the level of output and the hourly wage rate. Even so, all of the estimated coefficients in Table E-10 have the signs predicted in Table E-8.

The magnitude of the estimated parameter for the basic metal cost index is nearly three times larger than that of the estimated parameter for pounds of metal processed per establishment. Since the absolute values of the two confidence intervals for these two parameters do not overlap, it is possible to say that the basic metal cost index exerts more influence on unit cost than does the level

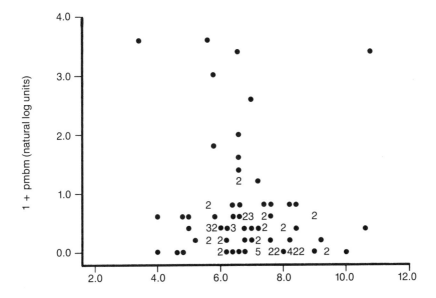

Pounds of metal processed (natural log units)

Figure E-4: Ratio of Processed Metal Costs to Basic Metal Costs versus Output: Pooled Four-Digit Data Set

of output. The correlation between the level of the basic metal cost index and the level of output is about the same as in the total data set.

Very few of the industries in SIC 34 purchase sizable quantities of processed metal inputs. While the estimate of the parameter for the degree of assembly, b_3, is large in magnitude, it is the least precise of the three parameter estimates.[7] Its joint confidence interval includes zero, indicating that the coefficient might be insignificant when the two other variables are included in the model. The lack of precision of this estimate does not appear to be due to multicollinearity. The variable pmbm is not strongly correlated with any other explanatory variable or with any linear combination of variables. Considering that the ratio of processed metal cost to basic metal cost is negligible for all but several industries in SIC 34, it is not surprising that the variable pmbm does not have a significant effect on unit cost.

While the estimate of the wage parameter has the predicted sign, it is not significantly different from zero; this is probably the result of a high correlation between the hourly wage rate and the output measure.

7. The ratio of processed metal cost to basic metal cost is zero for some industries in SIC 34, making it impossible to compute the natural log for the variable. To get around this problem, the constant 1 was added to the ratio and the log of (1 + pmbm) was computed.

E.2.3. Regression Results for SIC 35

The explanatory variables account for nearly 89 percent of the variation in unit cost in SIC 35, which has the largest number of observations of the four major groups. All of the parameter estimates have the signs predicted in Table E-8. The parameter estimates in Table E-11 are more precise than those in Table E-12, because of the lower degree of multicollinearity among the explanatory variables. The joint confidence intervals for the three parameters of interest in SIC 35 are shown in Table E-15.

The estimated magnitude of the parameter for the basic metal cost index is over four times that of the output elasticity. The basic metal index is positively correlated to the pounds of metal processed per establishment, although the correlation is not nearly as strong as in the total data set. The estimate for β_3 is positive and significant, indicating that unit cost tends to increase as the ratio of processed metal cost to basic metal cost increases, with all other factors held constant. This suggests that unit cost increases as the relative proportion of assembly to metal shaping increases.

The estimate of the wage parameter is positive and significant. Although the estimate of the coverage parameter has the predicted sign, it is negligible in magnitude and not significantly different from zero. This suggests that differences in material coverage within this group of industries do not account for variations in unit cost.

E.2.4. Regression Results for SIC 36

The explanatory variables in SIC 36 account for almost 94 percent of the variation in unit cost, which is the highest "goodness-of-fit" measure among the four major industry groups. The regression of pounds of metal against the other explanatory variables in the bottom half of Table E-12 indicates a strong degree of multicollinearity among the explanatory variables. Nonetheless, all of the parameter estimates have the sign predicted in Table E-8, and all are significantly different from zero, except for the coverage parameter. The simple correlation between pounds of metal processed per establishment and the basic metal cost index is the strongest among the four groups. There is a strong tendency for the industries processing low volumes of metal to use the more expensive basic metal inputs and for the industries processing high volumes of metal to use the less expensive basic metal inputs. This might be because of the fact that some of the small-volume producers in this industry (such as SIC 361, electric distribution equipment, and SIC 362, electrical industrial apparatus) make use of large amounts of copper wire and brass, which are expensive basic metal inputs.

The joint confidence interval for the output elasticity, β_1, is wide, because of the high degree of collinearity among the explanatory variables. Nonetheless, it does not include zero. The expected value of the magnitude of the basic metal cost index parameter is five times that of the output elasticity. While there is a high degree of collinearity among pounds of metal processed per establishment, the basic metal cost index, and the hourly wage rate, it still appears that unit cost decreases with the level of output and that the basic metal cost index exerts a greater influence on unit cost than does the level of output.

The confidence interval for β_3 indicates that this parameter is significant, and that an increase in the ratio of processed metal cost to basic metal cost tends to increase unit cost. As expected, industries with a greater degree of assembly have higher unit cost, with all other factors held constant.

E.2.5. REGRESSION RESULTS FOR SIC 37

There are only 14 observations in SIC 37, the smallest number of all four groups. The estimate of the sample standard deviation (designated by S in Tables E-9 through E-13) is the highest. The explanatory variables account for less than 80 percent of the variation in unit cost, which is less explanatory power than is achieved in the three other groups. All estimates have the signs predicted in Table E-8, with the one exception of b_5, the estimated parameter for material coverage. However, the small t-ratio for this variable indicates that it may be insignificant, given that the other variables are included in the model.

Pounds of metal processed per establishment and the basic metal index are negatively correlated, but not as strongly as in the other three major groups. The degree of multicollinearity among the explanatory variables is not very high. Nonetheless, the t-ratios for several of the coefficients indicate that they are insignificant, assuming the other variables are included in the model.

The expected value of the estimate of the output elasticity, $b_1 = -0.604$, is larger in absolute magnitude than for the other groups. In contrast with the results for the other groups, the confidence interval for the parameter for the basic metal cost index, β_2, and for the degree of assembly, β_3, includes zero, indicating that these parameters may be insignificant. This is the only group where the data analysis suggests that the level of output exerts more influence on unit cost than does the basic metal cost index.

E.2.6. COMBINING THE DATA SETS OF THE FOUR MAJOR GROUPS

Pooling subsets of observations from the same underlying population and increasing the total number of observations increase the precision of the parameter estimates. The distribution of the error terms, ϵ, is used to define the characteristics of the separate samples and to test if they are from the same underlying population. A key assumption in the normal error regression model is that the error terms are independent and normally distributed $N(0, \sigma)$, e.g., the error terms have a zero mean and a constant variance. An estimate of the variance of the error term, σ, is given by the error mean square (MSE), which is the sum of the squared error terms divided by its degrees of freedom. The error mean square is referred to as the residual mean square in the summary tables. It is given below for each of the four regressions.

Data Set	Error Mean Square (MSE)
SIC 34	0.05902
SIC 35	0.05085
SIC 36	0.06955
SIC 37	0.3729

What constitutes a significant difference among the error mean squares? One must choose between the two alternatives:

$$H_0: \sigma_{\text{regression } i} = \sigma_{\text{regression } j}$$
$$H_1: \sigma_{\text{regression } i} \neq \sigma_{\text{regression } j}$$

The test statistic for comparing the equality of error variances for two regression equations is given by

$$F* = \frac{MSE_{\text{regression equation } i}}{MSE_{\text{regression equation } j}}$$

where

MSE = error mean square

There is a $1 - a$ probability that the error variances of the two regressions are not significantly different if

$$F(a, df_i, df_j) < \frac{MSE_i}{MSE_j} < F(1 - a, df_i, df_j)$$

where

df_i = degrees of freedom of the error mean square for regression i

MSE_i = error mean square for regression i

Following this procedure, one would maintain, that with a $1 - a = 0.95$ confidence level, there is no significant difference among $\sigma_{\text{SIC 34}}$, $\sigma_{\text{SIC 35}}$, and $\sigma_{\text{SIC 36}}$, but there is a significant difference between $\sigma_{\text{SIC 37}}$ and the other three error variances. This raises some question about pooling the data sets from all four major groups together. The general linear test statistic is used to choose whether the results obtained from the four separate regressions are significantly different from the results obtained from the pooled data set. The test statistic is given by

$$F* = \frac{SSE(R) - SSE(F)}{df_e - df_f} / \frac{SSE(F)}{df_f}$$

where

SSE[F] = error sum of squares for the full model

SSE[R] = error sum of squares for the reduced model

df_f = degrees of freedom for the full model

df_r = degrees of freedom for the reduced model

full model = separate regressions for SIC 34, 35, 36, and 37

reduced model = one regression using the pooled data set

The decision rule is to conclude that the results of the reduced model (pooled regression) are not significantly different from the results of the full model (separate regressions) if

$$F^* < F(1 - a, df_r - df_f, df_f)$$

where
$F^* = 3.86$ and
$F(0.95, 18, 77) \sim 1.73$

Since F^* is larger than the value of the F statistic, it is concluded that the results of the pooled regression are significantly different from the results of the four separate regressions. This indicates that the observations from all four separate regressions do not come from the same underlying population. This casts some doubt on pooling the observations from SIC 37 with those from SIC 34, 35, and 36.[8]

The observations from all four major groups are pooled together in the combined data set, based on the justification that there is known to be a great deal of commonality in the types of operation performed across all of the industries in the metalworking sector, despite differences in the end use of the products. The assumptions of the normal error model for the combined data set were carefully checked, and there is no apparent violation of the assumptions.[9] While there are apparently some differences which distinguish the major groups of metalworking industries from one another, there also appears to be a strong degree of commonality among all of these industries, as evidenced by the good fit of the regression model and by the fact that the assumptions of the model do not appear to be violated.

8. It is noted that if the 14 observations from SIC 37 are excluded from the total data set, and if value added per pound is regressed against the five explanatory variables, then the expected value of the output elasticity is approximately −0.20 (as opposed to −0.295).

9. A plot of the residuals against the independent variable and a plot of the residuals against the fitted value of the independent variable were examined, and there was no indication that the variance of the error terms was not constant. Both a histogram of the residuals and a plot of the cumulative frequency distribution of the standardized residuals (the "normal scores") suggest they are normally distributed. A "runs test" of the residuals, as well as the plot of the residuals against the fitted value of the independent variable, suggests they are randomly distributed with zero mean.

Bibliography

Abraham, R. G., T. Csakvary, and G. Boothroyd. 1977. *Programmable assembly research technology transfer to industry, 5th bi-monthly report.* Technical Report NTIS No. PB285886, Westinghouse Research and Development Center, Pittsburgh, PA.

American Machinist. 1978. The 12th *American Machinist* Inventory of Metalworking Equipment, 1976–1978. *American Machinist* 122(12):133–148.

American Machinist. 1983. The 13th *American Machinist* Inventory of Metalworking Equipment, 1983. *American Machinist* 127(11):113–144.

American Society for Metals. 1967. *Machining.* Vol. 3. *The Metals Handbook,* 8th Ed. Metals Park, OH: American Society for Metals.

Anderson, R. H. 1972. Programmable Automation: The Bright Future of Computers in Manufacturing. *Datamation* 18:46–52.

Ardnt, G. 1977. Integrated Flexible Manufacturing Systems: Towards Automation in Batch Production. *New Zealand Engineering* 32(7):150–155.

Ayres, R. U., and S. M. Miller. 1982. Industrial Robots on the Line. *Technology Review* 85(4):35–45.

Ayres, R. U., and S. M. Miller. 1983. *Robotics: Applications and social implications.* Cambridge, MA: Ballinger Publishing.

Barash, M. M. 1979. Computerized Systems in the Scheme of Things. In *1979 fall industrial engineering conference,* 239–246. Norcrosse, GA: American Institute of Industrial Engineers.

Bolz, R. W. 1976. *Production Processes: The productivity handbook,* 5th ed. New York: Industrial Press.

Boothroyd, G., C. Poli, and L. E. Murch. 1982. *Manufacturing engineering and materials processing: Automatic assembly.* New York: Marcel Dekker.

Borzcik, P. S. 1980. Flexible Manufacturing Systems. In *Machine tool systems management and utilization,* ed. A. R. Thompson, 62–74. Machine Tool Task Force Report on the Technology of Machine Tools, Vol. 2. Livermore, CA: Livermore National Laboratory.

Bourne, D. A., and M. S. Fox. 1984. Autonomous Manufacturing: Automating the Job Shop. *IEEE Computer Magazine* 17(9):76–88.

Buffa, E. S. 1980. *Modern production operations management:* 6th ed. New York: John Wiley and Sons.

Bureau of the Census, Industry Division, U. S. Department of Commerce. 1981a. *1977 census of manufactures: General summary.* MC77-SR-1. Washington, DC: Government Printing Office.

Bureau of the Census, Industry Division, U. S. Department of Commerce. 1981b. *1977 census of manufactures: Selected materials consumed.* MC77-SR-11. Washington, DC: Government Printing Office.

Bureau of the Census, Industry Division, U. S. Department of Commerce. 1981c. *1977 census of manufactures, Vol. 2. Industry statistics,* Part 2. *SIC major groups, 27–34.* Washington, DC: Government Printing Office.

Bureau of the Census, Industry Division, U. S. Department of Commerce. 1981d. *1977 census of manufactures, Vol. 2. Industry statistics,* Part 3. *SIC major groups, 35–39.* Washington, DC: Government Printing Office.

Bureau of Labor Statistics, U. S. Department of Labor. 1980. *Occupational employment in manufacturing industries, 1977.* Bulletin 2057. Washington, DC: Government Printing Office.

Bureau of Labor Statistics, U.S. Department of Labor. 1982a. *The national OES-based industry-occupation matrix for 1980.* Washington, DC: Government Printing Office.

Bureau of Labor Statistics, U.S. Department of Labor. 1982b. *Occupational employment in manufacturing industries (1980).* Washington, DC: Government Printing Office.

Bylinski, G. 1983. The Race to the Automatic Factory. *Fortune* 107(4):52–64.

Carnegie-Mellon University. 1981. *The impacts of robotics on the workforce and workplace.* Pittsburgh: Department of Engineering and Public Policy, Carnegie-Mellon University. (A student project co-sponsored by the Department of Engineering and Public Policy, the School of Urban and Public Affairs, and the College of Humanities and Social Sciences.)

Carter, C. F., Jr. 1980. Machine Tool Characteristics Required for Versatility and Systems Use. In *Machine tool systems management and utilization,* ed. A. R. Thompson, 8.6-1 - 8.6-9. Machine Tool Task Force Report on the Technology of Machine Tools, Vol. 2. Livermore, CA: Lawrence Livermore National Laboratory.

Carter, C. F., Jr. 1982. Towards Flexible Automation. *Manufacturing Engineering* 89(2):75–79.

Ciborra, C., P. Migliarese, and P. Romano. 1980. Industrial Robots in Europe. *Industrial Robot* 7(3):164–167.

Cook, N. H. 1975. Computer-Managed Parts Manufacturing. *Scientific American* 232(2):22–29.

Cook, N. H., D. Gossard, R. Melville, and T. Ruegsegger. 1978. *Design and analysis of computerized manufacturing systems for small parts with emphasis on non-palletized parts of rotation.* Technical Report 3, Materials Processing Laboratory, Department of Mechanical Engineering, Massachusetts Institute of Technology. Prepared for the National Science Foundation, Washington, DC.

Dallas, D. B. 1981. Major Causes of Productivity Loss. *Manufacturing Engineering* 87(3):79–82.

David, P. A. 1982. Comments. In *Micro-electronics, robotics and jobs,* comp. Working Party on Information, Computer and Communications Policy, 146–157. Paris: OECD.

Draper Lab. 1983. *Flexible manufacturing system handbook.* Vol. 2, *Description of the technology.* Technical Report 12703, Charles Stark Draper Laboratory, Inc. Prepared for the U.S. Army Tank-Automotive Command, Warren, MI.

Dreyfoos, W. D., and P. F. Stregevsky. 1985. Robot Applications in Aerospace Manufacturing. In *Handbook of industrial robotics*, ed. S. Y. Nof, Ch. 41, 834–843. New York: John Wiley and Sons.

Eikonix Corporation. 1979. *Technology assessment: The impact of robots.* Technical Report. EC/2405801-FR-1. Final report for the National Science Foundation, Washington, DC, Grant No. ERS-76-00637.

Engelberger, J. F. 1980. *Robotics in practice.* New York: American Management Association.

Flora, P. C. 1982. *1982 robotics industry directory.* Lacanada, CA: Technical Data Base Corp.

Frost and Sullivan, Inc. 1974. *U. S. Industrial Robot Market.* Report. New York.

Funk, J. L. 1984. The potential societal benefits from developing flexible assembly technologies. Ph.D. diss., Carnegie-Mellon University, Pittsburgh.

Groover, M. P. 1980. *Automation, production systems and computer-aided manufacturing.* Englewood Cliffs, NJ: Prentice-Hall.

Ham, I. 1984. Group Technology Applications for Computer-integrated Manufacturing. *Robotics and Computer-Integrated Manufacturing* 1(^{3}/$_4$):231–235.

Hartley, J. 1985a. Yamazaki. In *Flexible manufacturing systems,* ed. H. J. Warnecke and R. Steinhilper, 271–279. Bedford, England: IFS Publications Ltd.

Hartley, J. 1985b. Fanuc. In *Flexible manufacturing systems,* ed. H. J. Warnecke and R. Steinhilper, 286. Bedford, England: IFS Publications Ltd.

Hasegawa, Y. 1985. Evaluation and Economic Justification. In *Handbook of industrial robotics,* ed. S. Y. Nof, Ch. 33, 665–687. New York: John Wiley and Sons.

Houtzeel, A. 1985. Process Planning and Group Technology. In *CAD/CAM handbook,* ed. T. Teicholz, Ch. 17, 17.1–17.27. New York: McGraw-Hill.

Hunt, H. A., and T. L. Hunt. 1982. *Robotics: Human resource implications for Michigan.* Technical Report, W. E. Upjohn Institute for Employment Research, Kalamazoo, MI. Prepared for the Michigan Occupational Information Coordinating Committee, Lansing.

Hunt, H. A., and T. L. Hunt. 1983. *Human resource implications of robotics.* Kalamazoo, MI: W. E. Upjohn Institute for Employment Research.

Hunt, T. L., and H. A. Hunt. 1985. An Assessment of Data Sources to Study the Employment Effects of Technological Change. In *Technology and employment effects: An interim report,* 1–116. Washington, DC: National Academy Press.

Industrial Relations Counselors, Inc. 1985. *IRC study of robotics on human resources and human relations.* Report. Industrial Relations Counselors, New York.

Industrial Robot. 1981. Robotics in the UK. *Industrial Robot* 8(1):32–38.

Jacobson, L., and R. Levy. 1983. *The effect of technical change on labor.* In *Automation and the Workplace.* comp. Office of Technology Assessment, 75–80. Washington, DC: U.S. Congress, Office of Technology Assessment.

Jaikumar, R. 1986. Postindustrial Manufacturing. *Harvard Business Review* 64(6):69–76.

Kaminski, M. A. 1986. Protocols for Communicating in the Factory. *IEEE Spectrum* 23(4):56–62.

Kaplan, R. S. 1983. Measuring Manufacturing Performance: A New Challenge for Managerial Accounting Research. *The Accounting Review* 58(4):686–705.

Kaplan, R. S. 1983. Measuring Manufacturing Performance: A New Challenge for Managerial Accounting Research. *The Accounting Review* 58(4):686–705.

Kaplan, R. S. 1986. Must CIM Be Justified by Faith Alone? *Harvard Business Review* 64(2): 87–95.

Kearney & Trecker Corporation, Special Products Division. 1982. *Manufacturing systems applications workbook.* Milwaukee: Kearney & Trecker Corp.

Kendrick, J. W. 1972. *Economic accounts and their use.* New York: McGraw-Hill.

Klahorst, H. T. 1983. How to Plan Your FMS. *Manufacturing Engineering* 91(3):52–54.

Klotz, T. 1984. The flexible manufacturing factory. Paper read at seminar, Flexible Manufacturing Systems. Sponsored by the Robot Institute of America, 26 January, Pittsburgh.

Krause, J. M. 1982. Robotics impact on human resources. Bachelor's thesis, General Motors Institute, Flint, MI.

Lamyai, T., Y. W. Rhee, and L. E. Westphal. 1978. *Factor substitution, returns to scale and the organization of production in the mechanical engineering industry.* Technical Report, Economics of Industry Department, the World Bank, Washington, DC.

Lamyai, T., Y. W. Rhee, and L. E. Westphal. 1982. *Economics of specialization as a source of technological change in the mechanical engineering industries.* Technical Report, Economics of Industry Department, the World Bank, Washington, DC.

Lary, H. R. 1968. *Imports of manufactures from less developed countries.* New York: National Bureau of Economic Research; distributed by Columbia University Press.

Leontief, W., and F. Duchin. 1984. *The impacts of automation on employment, 1963–2000.* Technical Report, Institute for Economic Analysis, New York University, New York. Final report to the National Science Foundation, Washington, DC. Contract No. PRA-8012844.

Lund, R. 1977. *Numerically controlled machine tools: A study of U.S. experience.* Technical Report CPA-78-2, Center for Policy Alternatives, Massachusetts Institute of Technology, Cambridge.

Machinability Data Center. 1980. *Machining data handbook.* Cincinnati: Metcut Research Associates.

Manufacturing Engineering. 1983. Special issue on flexible manufacturing systems. *Manufacturing Engineering.* 91(3).

Marshall, P. W. 1975. *Automation in die casting—a decade of experience.* Technical Report G-T75-022, Society of Die Casting Engineers, Detroit.

Mayer, J. E., and D. Lee. 1980a. Estimated Requirements for Machine Tools during the 1980–1990 Period. In *Machine tool systems management and utilization,* ed. A. R. Thompson, 31–34. Machine Tool Task Force Report on the Technology of Machine Tools, Vol. 2. Livermore, CA: Lawrence Livermore National Laboratory.

Mayer, J. E., and D. Lee. 1980b. Future Machine Tool Requirements for Achieving Increased Productivity. In *Machine tool systems management and utilization,* ed. A. R. Thompson, 8.4-1–8.4-12. Machine Tool Task Force Report on the Technology of Machine Tools, Vol.2. Livermore, CA: Lawrence Livermore National Laboratory.

Merchant, M. E. 1983. Current Status of, and Potential for, Automation in the Metalworking Manufacturing Industry. *Annals of the CIRP* 32(2):519–523.

Merchant, M. E. 1985. The Importance of Flexible Manufacturing Systems to the Realisation of Full Computer-Integrated Manufacturing. In *Flexible manufacturing systems,* ed. H. J. Warnecke and R. Steinhilper, 27–43. Bedford, England: IFS Publications Ltd.

Meyer, J. D. 1985. An Overview of Fabrication and Processing Applications. In *Handbook of industrial robotics*, ed. S. Y. Nof, Ch. 39, 807–820. New York: John Wiley and Sons.

Miller, S. M. 1983. Potential impacts of robotics on manufacturing costs within metalworking industries. Ph.D. diss., Carnegie-Mellon University, Pittsburgh.

Miller, S. M. 1985. Impacts of Robotics and Flexible Manufacturing Technologies on Manufacturing Costs and Employment. In *The management of technology and productivity in manufacturing*, ed. P. R. Kleindorfer, Ch. 3, 73–110. New York: Plenum Press.

Miller, S. M., and S. R. Bereiter. 1985. *Modernizing to computer-integrated production technologies in a vehicle assembly plant: Lessons for analysts and managers of technological change*. Technical Report, Graduate School of Industrial Administration, Carnegie-Mellon University, Pittsburgh.

Muther, R. 1971. Plant Layout. In *The industrial engineering handbook*, ed. H. B. Maynard, 11-26–11-62. New York: McGraw-Hill.

Nam, J. W., Y. W. Rhee, and L. E. Westphal. 1973. *Data development for a study of the scope of capital-labor substitution in the mechanical engineering industries*. Technical Report, Economics of Industry Division, the World Bank, Washington, DC.

National Research Council 1984. *Computer integration of engineering design and production: A national opportunity*. Washington, DC: National Academy Press.

Neter, J., and W. Wasserman. 1974. *Applied linear statistical models*. Homewood, IL: Richard D. Irwin, Inc.

Newsweek. 1981. Where the Jobs Are—and Aren't. *Newsweek* 98(21):88–90.

OECD. See Organization of Economic Cooperation and Development.

Office of Technology Assessment (OTA). U. S. Congress. 1983. *Automation and the workplace: Selected labor, education and training issues*. Technical Memorandum, U. S. Congress, Office of Technology Assessment, Washington, DC.

Office of Technology Assessment (OTA). U. S. Congress. 1984. *Computerized manufacturing automation: Employment, education, and the workplace*. Washington, DC: U. S. Government Printing Office.

Office of Technology Assessment (OTA). U. S. Congress. 1986. *Technology and structural unemployment: Reemploying displaced adults*. Washington, DC: U. S. Government Printing Office.

Organization of Economic Co-operation and Development (OECD). 1982. *Micro-electronics, robotics and jobs*, comp. Working Party on Information, Computer and Communications Policy. Paris: OECD.

Ostwald, P. F. 1981. *American machinist manufacturing cost estimating guide, 1982 ed.* New York: McGraw-Hill.

OTA. See Office of Technology Assessment.

Personick, V. A. 1981. The Outlook for Industry Output and Employment through 1990. *Monthly Labor Review* 104(8):28–41.

Porter, A. L., F. A. Rossini, J. Eshelman-Bell, D. D. Jenkins, and D. J. Cancelleri. 1985. Industrial Robots—A Strategic Forecast Using the Technological Delivery System Approach. *IEEE Transactions on Systems, Man, and Cybernetics* 15(4):521–527.

Productivity International, Inc. 1982. *A survey of industrial robots*. 2d ed. Dallas: Leading Edge Publications.

Roberts, A. D., and S. Lapidge. 1977. *Manufacturing processes*. New York: McGraw-Hill.

Robotic Industries Association. 1985. *3rd quarter 1985 statistical report*. Dearborn, MI: RIA.

Romero, C. J., S. Toye, and S. Baldwin. 1984. Summary of the Approach Used in "The Impacts of Automation on Employment, 1963–2000": A Report by Wassily Leontief and Faye Duchin. Prepared for the National Commission on Employment Policy, Washington, DC.

Rosegger, G. 1980. *The economics of production and innovation*. New York: Pergamon Press.

Sanderson, A. C., and G. Perry. 1983. Sensor-based Robotic Assembly Systems: Research and Applications in Electronic Manufacturing. *Proceedings of the IEEE* 71(7):856–871.

Sekiguchi, Y. 1985. Factory Automation in Japan. In *Innovations in management: The Japanese corporation,* ed. Y. Monden et al., Ch. 9, 99–144. Norcrosse, GA: Industrial Management Press.

Smith, D. N., and P. Heytler, Jr. 1985. *Industrial robots: Forecasts and trends*. 2d ed. Dearborn, MI: Society of Manufacturing Engineers.

Smith, D. N., and R. C. Wilson. 1982. *Industrial robots: A Delphi forecast of markets and technology*. Dearborn, MI: Society of Manufacturing Engineers.

Susnjara, K. 1982. *A manager's guide to industrial robots*. Shaker Heights, OH: Corinthian Press.

Tanner, W. R., and W. F. Adolfson. 1982. *Robotics use in motor vehicle manufacture*. Technical Report, submitted to the U. S. Department of Transportation, Washington, DC.

Taylor, T., Jr. 1979. *Automative manufacturing assessment system*. Vol. 4. *Engine manufacturing systems*. Technical Report DOT-TSC-NHTSA-79-29. IV, U.S. Department of Transportation, Transportation Systems Center, Cambridge, MA.

Tech Tran Corporation. 1983. *Industrial robots: A summary and forecast*. Naberville, IL: Tech Tran Corp.

Tepsic, R. M. 1983. How to Justify Your FMS. *Manufacturing Engineering* 91(3):50–52.

Thomson, A. R. 1980. Introduction and Summary. In *Machine tool systems management and utilization,* ed. A. R. Thompson, 1–26. Machine Tool Task Force Report on the Technology of Machine Tools, Vol. 2. Livermore, CA: Lawrence Livermore National Laboratory.

Toepperwein, L. L., and M. T. Blackman. 1980. *ICAM robotics application guide*. Technical Report AFWAL-TR-80-4042, Vol. 2, ICAM Program, Wright Patterson Air Force Base, Ohio. Also available from Noyes Data Corp., Park Ridge, NJ, under the title *Robotics Applications for Industry,* 1983.

U. S. Department of Commerce, Bureau of Economic Analysis. 1979. *The detailed input-output structure of the U.S. economy: 1972*. Washington, DC: Government Printing Office.

Varian, H. R. 1978. *Microeconomic analysis*. New York: W. W. Norton.

Vedder, R. K. 1982. *Robotics and the economy*. Technical Report, U.S. Congress, Joint Economic Committee. Washington, DC.

Volkholz, V. 1982. Trends in the Use of Industrial Robots in the 805—the Case of the Federal Republic of Germany. In *Micro-electronics, robotics, and jobs,* comp. Working Party on Information, Computer and Communications Policy, 173–195. Paris: OECD.

Warnecke, H. J., and R. Steinhilper. 1985. Flexible Manufacturing Systems and Cells. In *Flexible manufacturing systems,* ed. H. J. Warnecke and R. Steinhilper, 1–18. Bedford, England: IFS Publications Ltd.

Westphal, L. E., and Y. W. Rhee. 1982. The Methodology of Investment Analysis for a Non-Process Industry. Technical Report. Economics of Industry Division, the World Bank. Washington, DC.

Whitney, D. E., P. M. Catalano, T. L. DeFazio, R. E. Gustavson, A. S. Kondoleon, J. L. Nevins, and D. S. Seltzer. 1981. *Design and control of adaptable-programmable assembly systems.* Technical Report R-1406, Charles Stark Draper Laboratory, Inc. Prepared for the National Science Foundation, Washington, DC, Grant No. DAR77-23712.

Wright, P. K., D. A. Bourne, J. P. Colyer, J. A. E. Isasi, and G. S. Schatz. 1982. A Flexible Manufacturing Cell for Swaging. *Mechanical Engineering,* 104(10):76–83.

Yoshikowa, H., K. Rathmill, and J. Hatvany. 1981. *Computer-aided manufacturing: An international comparison.* Technical Report, sponsored by the National Research Council. National Academy Press, Washington, DC.

Index

Advanced Manufacturing Systems, 40
Aerospace manufacturing, 42, 46, 126
Aircraft and parts (SIC 3721), 219
Aluminum: castings, 211; shapes, 212
American Machinists Inventory of Metalworking Equipment (12th), 33, 35, 108, 109, 149; categories of metalcutting machines, 198, 199, 204
Apparel industry, 9
Assemblers, 27, 31, 32
Attrition rates, 62, 63–72; in the durable goods industries, 64
Automation, flexible. *See* FMSs
Automobile: industry, 22–23, 128, 132; stampings, 128, 132
AVCO-Williamsport (company), 157, 158
Ayres, Robert U., xxii–xxiii
Ayres and Miller study, xxii–xxiii, 17, 19, 24–26, 28–29; use of the engineering approach, 75

Basic metal cost: index (bmci), 163, 164, 165, 220, 232, 234; the ratio of processed metal cost to, 214–19, 236–39
Basic metal inputs, 210–14
Batch-production, 83–135, 137–42, 174–75; plants, 5, 6; and mass-production, 83–89, 174, 234–35; industries, identification of, 97–107; facilities, increasing output levels in, 150–67
Batch size, estimation of, 189–90. *See also* Output
BLS (Bureau of Labor Statistics), 21–22, 29; *Occupational Employment* survey, 27, 191, 203; projections of replacement needs to offset attrition, 63
Blue-collar occupations, 17, 18
Bolt, nut, rivet and washer industry (SIC 3452), 103, 129
Bonferroni method, 232

Calculating and accounting equipment (SIC 3574), 126, 219
CAM (computer-aided manufacturing technology), 5, 77, 136; and increases in the level of output, 141–42, 145, 147, 148, 150
Capital cost: per unit, and the volume of output, 84, 85–95, 107; calculation of, 93–94; the elasticity of, to labor cost, 95, 109–11
Carnegie Mellon Robotics Institute, xxii, 20, 171
Carnegie Mellon University robotics survey. *See* CMU survey
Caterpillar Tractor (company), 157, 158, 182
Census of Manufactures (Bureau of the Census), 5, 35–36, 53, 98, 99, 108; listings of the total dollar value of materials, 102; materials survey of the, 103–4, 219n5; overview of the SIC system in the, 179–80; accounting of interfirm transfers in, 218; survey of wage rates, 220
Cincinnati Milacron T3, 171
Classification, of industries by mode of production, 118–26; in the fabricated metal products industry, (SIC 34), 126; in the machinery industry, (SIC 35), 127; in the electrical and electronic machinery industry (SIC 36), 128; in the transportation industry (SIC 37), 129
CMU survey (Carnegie-Mellon University robotics survey), 7, 12–18, 22–23, 26; estimates of displacement by Level I robots, 14–17, 24, 33–34, 58, 203; estimates of displacement by Level II robots, 14–17, 33, 58, 60, 203; estimates for the potential for robot use, 33, 35; of motivations for installing robots, 42
Commerce, U.S. Department of: 1972 input-output table of the, 104; Office of Federal Statistical Policy and Standards, 179
Communication and control systems, 60, 61